ABSTRACT ALGEBRA

SECOND EDITION

ABSTRACT ALGEBRA

SECOND EDITION

I.N. Herstein
University of Chicago

Macmillan Publishing Company
New York
Collier Macmillan Publishers
London

Macmillan Publishing Company.
866 Third Avenue, New York, New York 10022

Collier Macmillan Canada, Inc.

Library of Congress Cataloging in Publication Data

Herstein, I. N.
   Abstract algebra / I. N. Herstein. — 2nd ed.
        p.   cm.
     ISBN: 0-02-353822-8
     1. Algebra, Abstract. I. Title.
QA162.H47   1990   89-28308
512'.02—dc20   CIP

To Biśka

PREFACE TO THE SECOND EDITION

When I was invited to prepare this Second Edition of Herstein's Abstract Algebra, I felt that it would be a mistake to undertake any major changes in such an attractive and well balanced treatment of the subject. This opinion was echoed by everyone with whom I conferred. So, no major changes have been made, and Herstein's inimitable style and choice of content remain virtually untouched.

At the same time, important minor changes thread their way throughout the Second Edition. Some fix minor errors and roughspots, and others expand and clarify the discussion and examples.

There are also two small changes in format. First, a Symbol List has been added, for readers to use when they confront a forgotten symbol or look ahead. Second, and more importantly, a few problems have been marked with an asterisk (*). These problems serve as a vehicle to introduce some concepts and simple arguments which support or relate in some interesting way to the discussion. As such, they should be read carefully.

I am grateful to several people whose very substantial contributions to this edition will help make this delightful book by Herstein even more enjoyable for the reader. I take this opportunity to thank Georgia Benkart, Barbara Cortzen and Lynne Small for their creative and highly useful suggestions.

David J. Winter

PREFACE TO THE FIRST EDITION

In the last half-century or so abstract algebra has become increasingly important not only in mathematics itself, but also in a variety of other disciplines. For instance, the importance of the results and concepts of abstract algebra play an ever more important role in physics, chemistry, and computer science, to cite a few such outside fields.

In mathematics itself abstract algebra plays a dual role: that of a unifying link between disparate parts of mathematics and that of a research subject with a highly active life of its own. It has been a fertile and rewarding research area both in the last 100 years and at the present moment. Some of the great accomplishments of our twentieth-century mathematics have been precisely in this area. Exciting results have been proved in group theory, commutative and noncommutative ring theory, Lie algebras, Jordan algebras, combinatorics, and a host of other parts of what is known as abstract algebra. A subject that was once regarded as esoteric has become considered as fairly down-to-earth for a large cross section of scholars.

The purpose of this book is twofold. For those readers who either want to go on to do research in mathematics or in some allied fields that use algebraic notions and methods, this book should serve as an introduction—and, we stress, only as an introduction—to this fascinating subject. For interested readers who want to learn what is going on in an engaging part of modern mathematics, this book could serve that purpose, as well as provide them with some highly usable tools to apply in the areas in which they are interested.

The choice of subject matter has been made with the objective of introducing readers to some of the fundamental algebraic systems that are both interesting and of wide use. Moreover, in each of these systems the aim has been to arrive at some significant results. There is little purpose served in studying some abstract object without seeing some nontrivial consequences of the study. We hope that we have achieved the goal of presenting interesting, applicable, and significant results in each of the systems we have chosen to discuss.

As the reader will soon see, there are many exercises in the book. They are often divided into three categories: easier, middle-level, and harder (with an occasional very hard). The purpose of these problems is to allow students to test their assimilation of the material, to challenge their mathematical ingenuity, to prepare the ground for material that is yet to come, and to be a means of developing mathematical insight, intuition, and techniques. Readers should not become discouraged if they do not manage to solve all the problems. The intent of many of the problems is that they be tried—even if not solved—for the pleasure (and frustration) of the reader. Some of the problems appear several times in the book. Trying to do the problems is undoubtedly the best way of going about learning the subject.

A *Student's Solution Manual*, containing detailed solutions to many of the problems, is available from the publisher. An *Instructor's Manual* is also available.

We have strived to present the material in the language and tone of a classroom lecture. Thus the presentation is somewhat chatty; we hope that this will put the readers at their ease. An attempt is made to give many and revealing examples of the various concepts discussed. Some of these examples are carried forward to be examples of other phenomena that come up. They are often referred to as the discussion progresses.

We feel that the book is self-contained, except in one section—the second last one of the book—where we make implicit use of the fact that a polynomial over the complex field has complex roots (this is the celebrated *Fundamental Theorem of Algebra* due to Gauss), and in the last section where we make use of a little of the calculus.

We are grateful to many people for their comments and suggestions on earlier drafts of the book. Many of the changes they suggested have been incorporated and should improve the readability of the book. We should like to express our special thanks to Professor Martin Isaacs for his highly useful comments.

We are also grateful to Fred Flowers for his usual superb job of typing the manuscript, and to Mr. Gary W. Ostedt of the Macmillan Company for his enthusiasm for the project and for bringing it to publication.

With this we wish all the readers a happy voyage on the mathematical journey they are about to undertake into this delightful and beautiful realm of abstract algebra.

I. N. H.

CONTENTS

CHAPTER 3

The Symmetric Group   129

CHAPTER 4

Ring Theory   147

CHAPTER 5

Fields   207

CHAPTER **6**

SYMBOL LIST

$a \in S$	a is an element of the set S, 4
$a \notin S$	a is not an element of the set S, 4
$S \subset T,\ T \supset S$	S is a subset of the set T, 4
$S = T$	The sets S and T are equal (have the same elements), 4
ϕ	The empty set, 4
$A \cup B$	The union of the sets A and B, 4
$A \cap B$	The intersection of the sets A and B, 5
$\{s \in S \mid s \text{ satisfies } P\}$	The subset of elements of S satisfying P, 5
$A - B$	The difference of the sets A and B, 5
$S - A$	For $A \subset S$, the complement of A in S, 5
$(a,\ b)$	Ordered pair consisting of a, b (see also below), 6
$A \times B$	The Cartesian product of A and B, 6
$\mathbb{R}$	The set of real numbers, 9
$f\!: S \to T$	Function from the set S to the set T, 9
$f(s)$	Image of the element s under the function f, 9

$i: S \to S,\ i_S$	The identity function on S, 10
$f^{-1}(t)$	Inverse image to t under f, 12
$f^{-1}(A)$	Inverse image to a subset A of T under $f: S \to T$, 12
$f \circ g,\ fg$	Composition or product of functions f and g, 13
$A(S)$	Set of 1-1 mapping from a set S to S, 18
S_n	Symmetric group of degree n, 18
$n!$	n factorial, 19
$\mathbf{Z}$	Set of integers, 25
$m\mid n$	m divides n, 26
$m \nmid n$	m does not divide n, 26
$(a,\ b)$	Greatest common divisor of a, b (see also above), 27
$\mathbb{C}$	Set of complex numbers, 37
$i,\ -i$	Square roots of -1, 37
$z = a + bi$	Complex number z with real part a and imaginary part b, 37
$\bar{z} = a - bi$	Conjugate of complex number $z = a + bi$, 37
$1/z$	Inverse of the complex number z, 38
$\lvert z \rvert$	Absolute value of complex number z, 39
θ_n	Primitive n th root of unity, 43
$\mathbf{Q}$	Set of rational numbers, 49
E_n	Group of n th roots of unity, 49
$\lvert G \rvert$	Order of the group, G, 50
$C(a)$	Centralizer of a in G, 62
(a)	Cyclic group generated by a, 62
$Z(G)$	Center of group G, 63
$a \sim b$	a is equivalent to b in a specified sense, 67, 68
$a \equiv b \bmod(n)$	a is congruent to b modulo n (long form), 68
$a \equiv b(n)$	a is congruent to b modulo n (short form), 68
$[a]$	Class of all b equivalent to a, 69, 70
$\mathrm{cl}(a))$	Conjugacy class of a, 69, 70
$o(a))$	Order of elements a of a group, 70
$i_G(H))$	Index of H in G, 70
Z_n	Set of integers mod n, 72
U_n	Group of invertible elements of Z_n, 73
$\varphi(n)$	Euler function (phi function), 73

Ker (φ)	Kernel of the homomorphism p, 83
$N \triangleleft G$	N is a normal subgroup of G, 85
G/N	Quotient of a group G by a subgroup, N, 93
AB	Product of subsets A, B of a group, 95
$G_1 \times G_2 \times \cdots \times G_n$	Direct product of $G_1, G_2, \ldots, G_n$, 110
$\begin{pmatrix} a & b & \cdots & c \\ u & v & & w \end{pmatrix}$	Permutation sending a to u, b to v, $\ldots$, c to w, 130, 131
$(a, b, \ldots c)$	Cycle sending a to b, $\ldots$, c to a, 132
A_n	Alternating group of degree n, 142, 143
$\alpha_0 + \alpha_1 i + \alpha_2 j + \alpha_3 k$	Quaternion, 155
(a)	Ideal generated by a in a commutative ring, 171
$F[x]$	Polynomial ring over the field F, 179
$(g(x))$	Ideal generated by g (x) in a polynomial ring, 184
$R[x]$	Polynomial ring over ring R, 191
$v \in V$	Vector v in a vector space V, 212
αv	Scalar α times vector v, 212
$\alpha_1 v_1 + \cdots + \alpha_n v_n$	Linear combination of vectors $v_1, \ldots, v_n$, 214, 218
$\dim_F (V)$	Dimension of V over F, 218, 220
$U + V$	Sum of subspaces U, W of V, 225
$[K:F]$	Dimension of K over F, 226
$F[a]$	Ring generated by a over F, 231
$F(a)$	Field extension obtained by adjoining a to F, 232
$E(K)$	Field of algebraic elements of K over F, 234
$\phi_n(x)$	n th cyclotomic polynomial, 273

CHAPTER 1

Things Familiar and Less Familiar

1. A FEW PRELIMINARY REMARKS

For many readers this book will be their first contact with abstract mathematics. The subject to be discussed is usually called "abstract algebra," but the difficulties that the reader may encounter are not so much due to the "algebra" part as they are to the "abstract" part.

On seeing some area of abstract mathematics for the first time, be it in analysis, topology, or what-not, there seems to be a common reaction for the novice. This can best be described by a feeling of being adrift, of not having something solid to hang on to. This is not too surprising, for while many of the ideas are fundamentally quite simple, they are subtle and seem to elude one's grasp the first time around. One way to mitigate this feeling of limbo, or asking oneself "What is the point of all this?," is to take the concept at hand and see what it says in particular cases. In other words, the best road to good understanding of the notions introduced is to look at examples. This is true in all of mathematics, but it is particularly true for the subject matter of abstract algebra.

Can one, with a few strokes, quickly describe the essence, purpose, and background for the material we shall study? Let's give it a try.

We start with some collection of objects S and endow this collection with an algebraic structure by assuming that we can combine, in one or several ways (usually two), elements of this set S to obtain, once more, elements of this set S. These ways of combining elements of S we call *operations* on S. Then we try to condition or regulate the nature of S by imposing certain rules on how these operations behave on S. These rules are usually called the *axioms* defining the particular structure on S. These axioms are for us to define, but the choice made comes, historically in mathematics, from noticing that there are many concrete mathematical systems that satisfy these rules or axioms. We shall study some of the basic axiomatic algebraic systems in this book, namely *groups*, *rings*, and *fields*.

Of course, one could try many sets of axioms to define new structures. What would we require of such a structure? Certainly we would want that the axioms be consistent, that is, that we should not be led to some nonsensical contradiction computing within the framework of the allowable things the axioms permit us to do. But that is not enough. We can easily set up such algebraic structures by imposing a set of rules on a set S that lead to a pathological or weird system. Furthermore, there may be very few examples of something obeying the rules we have laid down.

Time has shown that certain structures defined by "axioms" play an important role in mathematics (and other areas as well) and that certain others are of no interest. The ones we mentioned earlier, namely groups, rings, and fields, have stood the test of time.

A word about the use of "axioms." In everyday language "axiom" means a self-evident truth. But we are not using everyday language; we are dealing with mathematics. An axiom is not a universal truth—whatever that may be—but one of several rules spelling out a given mathematical structure. The axiom is true in the system we are studying because we have forced it to be true by hypothesis. It is a license, in the particular structure, to do certain things.

We return to something we said earlier about the reaction that many students have on their first encounter with this kind of algebra, namely a lack of feeling that the material is something they can get their teeth into. Do not be discouraged if the initial exposure leaves you in a bit of a fog. Stick with it, try to understand what a given concept says, and most important, look at particular, concrete examples of the concept under discussion.

Problems

1. Let S be a set having an operation $*$ which assigns an element $a * b$ of S for any $a,\, b \in S$. Let us assume that the following two rules hold:

    1. If a, b are any objects in S, then $a * b = a$.

    2. If a, b are any objects in S, then $a * b = b * a$.

Show that S can have at most one object.

2. Let S be the set of all integers $0, \pm 1, \pm 2, \ldots, \pm n, \ldots$. For a, b in S define $*$ by $a * b = a - b$. Verify the following:

(a) $a * b \neq b * a$ unless $a = b$.

(b) $(a * b) * c \neq a * (b * c)$ in general. Under what conditions on a, b, c is $(a * b) * c = a * (b * c)$?

(c) The integer 0 has the property that $a * 0 = a$ for every a in S.

(d) For a in S, $a * a = 0$.

3. Let S consist of the two objects $\square$ and $\triangle$. We define the operation $*$ on S by subjecting $\square$ and $\triangle$ to the following conditions:

    1. $\square * \triangle = \triangle = \triangle * \square$.

    2. $\square * \square = \square$.

    3. $\triangle * \triangle = \square$.

Verify by explicit calculation that if a, b are any elements of S (i.e., a, b and c can be any of $\square$ or $\triangle$), then:

(a) $a * b$ is in S.

(b) $(a * b) * c = a * (b * c)$.

(c) $a * b = b * a$.

(d) There is a particular a in S such that $a * b = b * a = b$ for *all* b in S.

(e) Given b in S, then $b * b = a$, where a is the particular element in Part (d).

2.　SET THEORY

With the changes in the mathematics curriculum in the schools in the United States, many college students have had some exposure to set theory. This introduction to set theory in the schools usually includes the elementary notions and operations with sets. Going on the assumption that many readers will have some acquaintance with set theory, we shall give a rapid survey of those parts of set theory that we shall need in what follows.

First, however, we need some notation. To avoid the endless repetition of certain phrases, we introduce a shorthand for these phrases. Let S be a collection of objects; the objects of S we call the *elements* of S. To denote that a given element, a, is an element of S, we write $a \in S$—this is read "a is an element of S." To denote the contrary, namely that an object a is *not* an element of S, we write $a \notin S$. So, for instance, if S denotes the set of all positive integers $1, 2, 3, \ldots, n, \ldots$, then $165 \in S$, whereas $-13 \notin S$.

We often want to know or prove that given two sets S and T, one of these is a part of the other. We say that S is a *subset* of T, which we write $S \subset T$ (read "S is contained in T") if every element of S is an element of T. In terms of the notation we now have: $S \subset T$ if $s \in S$ implies that $s \in T$. We can also denote this by writing $T \supset S$, read "T contains S." (This does not exclude the possibility that $S = T$, that is, that S and T have exactly the same set of elements.) Thus, if T is the set of all positive integers and S is the set of all positive even integers, then $S \subset T$, and S is a subset of T. In the definition given above, $S \supset S$ for any set S; that is, S is always a subset of itself.

We shall frequently face the problem that two sets S and T, defined perhaps in distinct ways, are equal, that is, they consist of the same set of elements. The usual strategy for proving this is to show that both $S \subset T$ and $T \subset S$. For instance, if S is the set of all positive integers having 6 as a factor and T is the set of all positive integers having both 2 and 3 as factors, then $S = T$. (Prove!)

The need also arises for a very peculiar set, namely one having no elements. This set is called the *null* or *empty* set and is denoted by $\varnothing$; $\varnothing$ has the property that it is a subset of *any* set S.

Let A, B be subsets of a given set S. We now introduce methods of constructing other subsets of S from A and B. The first of these is the *union* of A and B, written $A \cup B$, which is defined: $A \cup B$ is that subset of S consisting of those elements of S that are elements of A *or* are elements of B. The "or" we have just used is somewhat different in meaning from the ordinary usage of the word. Here we mean that an element c is in $A \cup B$ if it is in A, or is in B, or is in *both*. The "or" is not meant to exclude the possibility that both things are true. Consequently, for instance, $A \cup A = A$.

If $A = \{1, 2, 3\}$ and $B = \{2, 4, 6, 10\}$, then $A \cup B = \{1, 2, 3, 4, 6, 10\}$.

We now proceed to our second way of constructing new sets from old. Again let A and B be subsets of a set S; by the *intersection* of A and B, written $A \cap B$, we shall mean the subset of S consisting of those elements that are both in A *and* in B. Thus, in the example above, $A \cap B = \{2\}$. It should be clear from the definitions involved that $A \cap B \subset A$ and $A \cap B \subset B$. Particular examples of intersections that hold universally are: $A \cap A = A$, $A \cap S = A$, $A \cap \varnothing = \varnothing$.

This is an opportune moment to introduce a notational device that will be used time after time. Given a set S, we shall often be called on to describe the subset, A, of S that satisfies a certain property P. We shall write this as $A = \{s \in S | s \text{ satisfies } P\}$. For instance, if A, B are subsets of S, then $A \cup B = \{s \in S | s \in A \text{ or } s \in B\}$ while $A \cap B = \{s \in S | s \in A \text{ and } s \in B\}$.

Although the notions of union and intersection of subsets of S have been defined for two subsets, it is clear how one can define the union and intersection of any number of subsets.

We now introduce a third operation we can perform on sets, the *difference* of two sets. If A, B are subsets of S, we define $A - B = \{a \in A | a \notin B\}$. So if A is the set of all positive integers and B is the set of all even integers, then $A - B$ is the set of all positive odd integers. In the particular case when A is a subset of S, the difference $S - A$ is called the *complement* of A in S and is written A'.

We represent these three operations pictorially. If A is Ⓐ and B is Ⓑ, then

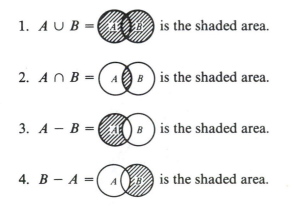

1. $A \cup B = $ is the shaded area.

2. $A \cap B = $ is the shaded area.

3. $A - B = $ is the shaded area.

4. $B - A = $ is the shaded area.

Note the relation among the three operations, namely the relation $A \cup B = (A \cap B) \cup (A - B) \cup (B - A)$. As an illustration of how one goes about proving the equality of sets constructed by such set-theoretic constructions, we prove this latter alleged equality. We first show that $(A \cap B) \cup (A - B) \cup (B - A) \subset A \cup B$; this part is easy for, by definition, $A \cap B \subset A$, $A - B \subset A$, and $B - A \subset B$, hence

$$(A \cap B) \cup (A - B) \cup (B - A) \subset A \cup A \cup B = A \cup B.$$

Now for the other direction, namely that $A \cup B \subset (A \cap B) \cup (A - B) \cup (B - A)$. Given $u \in A \cup B$, if $u \in A$ and $u \in B$, then $u \in A \cap B$, so it is certainly in $(A \cap B) \cup (A - B) \cup (B - A)$. On the other hand, if $u \in A$ but $u \notin B$, then, by the very definition of $A - B$, $u \in A - B$, so again it is certainly in $(A \cap B) \cup (A - B) \cup (B - A)$. Finally, if $u \in B$ but $u \notin A$, then $u \in B - A$, so again it is in $(A \cap B) \cup (A - B) \cup (B - A)$. We have thus covered all the possibilities and have shown that $A \cup B \subset (A \cap B) \cup (A - B) \cup (B - A)$. Having the two opposite containing relations of $A \cup B$ and $(A \cap B) \cup (A - B) \cup (B - A)$, we obtain the desired equality of these two sets.

We close this brief review of set theory with yet another construction we can carry out on sets. This is the *Cartesian product* defined for the two sets A, B by $A \times B = \{(a, b) | a \in A, b \in B\}$, where we *declare* the ordered pair (a, b) to be equal to the ordered pair (a_1, b_1) if and only if $a = a_1$ and $b = b_1$. Here, too, we need not restrict ourselves to two sets; for instance, we can define, for sets A, B, C, their Cartesian product as the set of ordered triples (a, b, c), where $a \in A$, $b \in B$, $c \in C$ and where equality of two ordered triples is defined component-wise.

Problems

Easier Problems

1. Describe the following sets verbally.
   (a) $S = \{$Mercury, Venus, Earth, $\ldots$, Pluto$\}$.
   (b) $S = \{$Alabama, Alaska, $\ldots$, Wyoming$\}$.

2. Describe the following sets verbally.
   (a) $S = \{2, 4, 6, 8, \ldots\}$.
   (b) $S = \{2, 4, 8, 16, 32, \ldots\}$.
   (c) $S = \{1, 4, 9, 16, 25, 36, \ldots\}$.

3. If A is the set of all residents of the United States, B the set of all Canadian citizens, and C the set of all women in the world, describe the sets $A \cap B \cap C$, $A - B$, $A - C$, $C - A$ verbally.

4. If $A = \{1, 4, 7, a\}$ and $B = \{3, 4, 9, 11\}$ and you have been told that $A \cap B = \{4, 9\}$, what must a be?

5. If $A \subset B$ and $B \subset C$, prove that $A \subset C$.

6. If $A \subset B$, prove that $A \cup C \subset B \cup C$ for any set C.

7. Show that $A \cup B = B \cup A$ and $A \cap B = B \cap A$.

8. Prove that $(A - B) \cup (B - A) = (A \cup B) - (A \cap B)$. What does this look like pictorially?

9. Prove that $A \cap (B \cup C) = (A \cap B) \cup (A \cap C)$.

10. Prove that $A \cup (B \cap C) = (A \cup B) \cap (A \cup C)$.

11. Write down all the subsets of $S = \{1, 2, 3, 4\}$.

Middle-Level Problems

*12. If C is a subset of S, let C' denote the complement of C in S. Prove the *De Morgan Rules* for A, B subsets of S, namely:
(a) $(A \cap B)' = A' \cup B'$.
(b) $(A \cup B)' = A' \cap B'$.

*13. Let S be a set. For any two subsets of S we define

$$A + B = (A - B) \cup (B - A) \qquad \text{and} \qquad A \cdot B = A \cap B.$$

Prove that:
(a) $A + B = B + A$.
(b) $A + \varnothing = A$.
(c) $A \cdot A = A$.
(d) $A + A = \varnothing$.
(e) $A + (B + C) = (A + B) + C$.
(f) If $A + B = A + C$, then $B = C$.
(g) $A \cdot (B + C) = A \cdot B + A \cdot C$.

*14. If C is a finite set, let $m(C)$ denote the number of elements in C. If A, B are finite sets, prove that

$$m(A \cup B) = m(A) + m(B) - m(A \cap B).$$

15. For three finite sets A, B, C find a formula for $m(A \cup B \cup C)$. (**Hint:** First consider $D = B \cup C$ and use the result of Problem 14.)

16. Take a shot at finding $m(A_1 \cup A_2 \cup \cdots \cup A_n)$ for n finite sets $A_1, A_2, \ldots, A_n$.

17. Use the result of Problem 14 to show that if 80% of all Americans have gone to high school and 70% of all Americans read a daily newspaper, then *at least* 50% of Americans have both gone to high school and read a daily newspaper.

18. A public opinion poll shows that 93% of the population agreed with the government on the first decision, 84% on the second, and 74% on the third, for three decisions made by the government. At least what percentage of the population agreed with the government on all three decisions? (**Hint:** Use the results of Problem 15.)

19. In his book *A Tangled Tale*, Lewis Carroll proposed the following riddle about a group of disabled veterans: "Say that 70% have lost an eye, 75% an ear, 80% an arm, 85% a leg. What percentage, *at least*, must have lost all four?" Solve Lewis Carroll's problem.

*20. Show, for finite sets A, B, that $m(A \times B) = m(A)m(B)$.

21. If S is a set having five elements:
    (a) How many subsets does S have?
    (b) How many subsets having four elements does S have?
    (c) How many subsets having two elements does S have?

Harder Problems

22. (a) Show that a set having n elements has 2^n subsets.
    (b) If $0 < m < n$, how many subsets are there that have exactly m elements?

3. MAPPINGS

One of the truly universal concepts that runs through almost every phase of mathematics is that of a *function* or *mapping* from one set to another. One could safely say that there is no part of mathematics where the notion does not arise or play a central role. The definition of a function from one set to another can be given in a formal way in terms of a subset of the Cartesian product of these sets. Instead, here, we shall give an informal and admittedly nonrigorous definition of a mapping (function) from one set to another.

Let S, T be sets; a *function* or *mapping f* from S to T is a *rule* that assigns to *each* element $s \in S$ a *unique* element $t \in T$. Let's explain a little more thoroughly what this means. If s is a given element of S, then there is *only one* element t in T that is associated to s by the mapping. As s varies over S, the t varies over T (in a manner depending on s). Note that by the definition given, the following is *not* a mapping. Let S be the set of all people in the world and T the set of all countries in the world. Let f be the rule that assigns to every person his or her country of citizenship. Then f is not a mapping from S to T. Why not? Because there are people in the world that enjoy a dual citizenship; for such people there would not be a *unique* country of citizenship. Thus, if Mary Jones is both an English and French citizen, f would not make sense, as a mapping, when applied to Mary Jones. On the other hand, the rule $f: \mathbb{R} \to \mathbb{R}$, where $\mathbb{R}$ is the set of real numbers, defined by $f(a) = a^2$ for $a \in \mathbb{R}$, is a perfectly good function from $\mathbb{R}$ to $\mathbb{R}$. It should be noted that $f(-2) = (-2)^2 = 4 = f(2)$, and $f(-a) = f(a)$ for all $a \in \mathbb{R}$.

We denote that f is a mapping from S to T by $f: S \to T$ and for the $t \in T$ mentioned above we write $t = f(s)$; we call t the *image* of s under f.

The concept is hardly a new one for any of us. Since grade school we have constantly encountered mappings and functions, often in the form of formulas. But mappings need not be restricted to sets of numbers. As we see below, they can occur in any area.

EXAMPLES

1. Let $S = \{$all men who have ever lived$\}$ and $T = \{$all women who have ever lived$\}$. Define $f: S \to T$ by $f(s) =$ mother of s. Therefore, f(John F. Kennedy) = Rose Kennedy, and according to our definition, Rose Kennedy is the image under f of John F. Kennedy.

2. Let $S = \{$all citizens of the United States$\}$ and $T = \{$positive integers$\}$. Define, for $s \in S, f(s)$ by $f(s) =$ age of s. This f defines a mapping from S to T.

3. Let S be the set of all objects for sale in a grocery store and let $T=$ {all real numbers}. Define $f: S \to T$ by $f(s) =$ price of s. This defines a mapping from S to T.

4. Let S be the set of all integers and let $T = S$. Define $f: S \to T$ by $f(m) = 2m$ for any integer m. Thus the image of 6 under this mapping, $f(6)$, is given by $f(6) = 2 \cdot 6 = 12$, while that of -3, $f(-3)$, is given by $f(-3) = 2(-3) = -6$. If $s_1, s_2 \in S$ are in S and $f(s_1) = f(s_2)$, what can you say about s_1 and s_2?

5. Let $S = T$ be the set of all real numbers; define $f: S \to T$ by $f(s) = s^2$. Does every element of T come up as an image of some $s \in S$? If not, how would you describe the set of all images $\{ f(s) | s \in S \}$? When is $f(s_1) = f(s_2)$?

6. Let $S = T$ be the set of all real numbers; define $f: S \to T$ by $f(s) = s^3$. This is a function from S to T. What can you say about $\{ f(s) | s \in S \}$? When is $f(s_1) = f(s_2)$?

7. Let T be any nonempty set and let $S = T \times T$, the Cartesian product of T with itself. Define $f: T \times T \to T$ by $f(t_1, t_2) = t_1$. This mapping from $T \times T$ to T is called the *projection* of $T \times T$ onto its first component.

8. Let S be the set of all positive integers and let T be the set of all positive rational numbers. Define $f: S \times S \to T$ by $f((m, n)) = m/n$. This defines a mapping from $S \times S$ to T. Note that $f((1, 2)) = \frac{1}{2}$ while $f((3, 6)) = \frac{3}{6} = \frac{1}{2} = f((1, 2))$, although $(1, 2) \neq (3, 6)$. Describe the subset of $S \times S$ such that $f((a, b)) = \frac{1}{2}$.

The mappings to be defined in Examples 9 and 10 are mappings that occur for any nonempty sets and play a special role.

9. Let S, T be nonempty sets, and let t_0 be a fixed element of T. Define $f: S \to T$ by $f(s) = t_0$ for every $s \in S$; f is called a *constant* function from S to T.

10. Let S be any nonempty set and define $i: S \to S$ by $i(s) = s$ for every $s \in S$. We call this function of S to itself the *identity function* (or *identity mapping*) on S. We may, at times, denote it by i_S (and later in the book, by e).

Now that we have the notion of a mapping we need some way of identifying when two mappings from one set to another are equal. This is not God given; it is for us to decide how to declare $f = g$ where $f: S \to T$ and $g: S \to T$. What is more natural than to define this equality via the actions of f and g on the elements of S? More precisely, we declare that $f = g$ if and only if $f(s) = g(s)$ for *every* $s \in S$. If S is the set of all real numbers and f is defined on S by $f(s) = s^2 + 2s + 1$, while g is defined on S by $g(s) = (s + 1)^2$, our definition of the equality of f and g is merely a statement of the familiar identity $(s + 1)^2 = s^2 + 2s + 1$.

Having made the definition of equality of two mappings, we now want to single out certain types of mappings by the way they behave.

> **Definition.** The mapping $f: S \to T$ is *onto* or *surjective* if every $t \in T$ is the image under f of some $s \in S$; that is, if and only if, given $t \in T$, there exists an $s \in S$ such that $t = f(s)$.

In the examples we gave earlier, in Example 1 the mapping is not onto, since not every woman that ever lived was the mother of a male child. Similarly, in Example 2 the mapping is not onto, for not every positive integer is the Social Security number of some U.S. citizen. The mapping in Example 4 fails to be onto because not every integer is even; and in Example 5, again, the mapping is not onto, for the number -1, for instance, is not the square of any real number. However, the mapping in Example 6 is onto because every real number has a unique real cube root. The reader can decide whether or not the given mappings are onto in the other examples.

If we define $f(S) = \{ f(s) \in S | s \in S \}$, another way of saying that the mapping $f: S \to T$ is onto by saying that $f(S) = T$.

Another specific type of mapping plays an important and particular role in what follows.

> **Definition.** A mapping $f: S \to T$ is said to be *one-to-one* (written 1-1) or *injective* if for $s_1 \neq s_2$ in S, $f(s_1) \neq f(s_2)$ in T. Equivalently, f is 1-1 if $f(s_1) = f(s_2)$ implies that $s_1 = s_2$.

In other words, a mapping is 1-1 if it takes distinct objects into distinct images. In the examples of mapping we gave earlier, in Example 1 the mapping is not 1-1, since two brothers would have the same mother. However in Example 2 the mapping is 1-1 because distinct U.S. citizens have distinct Social Security numbers (provided that there is no goof-up in Washington, which is unlikely). The reader should check if the various other examples given of mappings are 1-1.

Given a mapping $f: S \to T$ and a subset $A \subset T$, we may want to look at $B = \{s \in S | f(s) \in A\}$; we use the notation $f^{-1}(A)$ for this set B, and call $f^{-1}(A)$ the *inverse image of A under f*. Of particular interest is $f^{-1}(t)$, the inverse image of the subset $\{t\}$ of T consisting of the element $t \in T$ alone. If the inverse image of $\{t\}$ consists of only one element, say $s \in S$, we could try to define $f^{-1}(t)$ by defining $f^{-1}(t) = s$. As we note now, this need not be a mapping from T to S, but is so if f is 1-1 and onto. We shall use the same notation f^{-1} in cases of both subsets and elements. This f^{-1} does *not* in general define a mapping from T to S for several reasons. First, if f is not onto, then there is some t in T which is not the image of any element s, so $f^{-1}(t) = \varnothing$. Second, if f is not 1-1, then for some $t \in T$ there are at least two distinct $s_1 \neq s_2$ in S such that $f(s_1) = t = f(s_2)$. So $f^{-1}(t)$ is *not* a unique element of S—something we require in our definition of mapping. However, if f is both 1-1 and onto T, then f^{-1} indeed defines a mapping of T onto S. (Verify!) This brings us to a very important class of mappings.

Definition. The mapping $f: S \to T$ is said to be a 1-1 *correspondence* or *bijection* if f is both 1-1 and onto.

Now that we have the notion of a mapping and have singled out various types of mappings, we might very well ask: "Good and well, but what can we do with them?" As we shall see in a moment, we can introduce an operation of combining mappings in certain circumstances.

Consider the situation $g: S \to T$ and $f: T \to U$. Given an element $s \in S$, then g sends it into the element $g(s)$ in T; so $g(s)$ is ripe for being acted on by f. Thus we get an element $f(g(s)) \in U$. We claim that this procedure provides us with a mapping from S to U. (Verify!)

We define this more formally in the

> **Definition.** If $g: S \to T$ and $f: T \to U$, then the *composition*
> (or *product*), denoted by $f \circ g$, is the mapping $f \circ g: S \to U$
> defined by $(f \circ g)(s) = f(g(s))$ for every $s \in S$.

Note that to compose the two mappings f and g—that is, for $f \circ g$ to
have any sense—the *terminal set*, T, for the mapping g *must be* the
initial set, for the mapping f. *One special time when we can always
compose any two mappings is when $S = T = U$, that is, when we map S
into itself.* Although special, this case is of the utmost importance.
  We verify a few properties of this composition of mappings.

> **Lemma 1.3.1.** If $h: S \to T$, $g: T \to U$, and $f: U \to V$, then
> $f \circ (g \circ h) = (f \circ g) \circ h$.

Proof. How shall we go about proving this lemma? To verify that two
mappings are equal, we merely must check that they do the same thing
to every element. Note first of all that both $f \circ (g \circ h)$ and $(f \circ g) \circ h$
define mappings from S to V, so it makes sense to speak about their
possible equality.
  Our task, then, is to show that for every $s \in S$, $(f \circ (g \circ h))(s) =$
$((f \circ g) \circ h)(s)$. We apply the definition of composition to see that

$$(f \circ (g \circ h))(s) = f((g \circ h)(s)) = f(g\,(h(s))).$$

Unraveling
$$((f \circ g) \circ h)\,(s) = (f \circ g)\,(h(s)) = f(g(h(s))),$$
we do indeed see that

$$(f \circ (g \circ h))\,(s) = ((f \circ g) \circ h)(s)$$

for every $s \in S$. Consequently, by definition, $f \circ (g \circ h) = (f \circ g) \circ h$. $\square$

  (The symbol $\square$ will always indicate that the proof has been completed.)
  This equality is described by saying that mappings, under composition, satisfy the *associative law*. Because of the equality involved there is
really no need for parentheses, so we write $f \circ (g \circ h)$ as $f \circ g \circ h$.

Lemma 1.3.2. If $g: S \to T$ and $f: T \to U$ are both 1-1, then $f \circ g: S \to U$ is also 1-1.

Proof. Suppose that $(f \circ g)(s_1) = (f \circ g)(s_2)$; thus, by definition, $f(g(s_1)) = f(g(s_2))$. Since f is 1-1, we get from this that $g(s_1) = g(s_2)$; however, g is also 1-1, thus $s_1 = s_2$ follows. Since $(f \circ g)(s_1) = (f \circ g)(s_2)$ forces $s_1 = s_2$, the mapping $f \circ g$ is 1-1. $\square$

We leave the proof of the next Remark to the reader.

REMARK. If $g: S \to T$ and $f: T \to U$ are both onto, then $f \circ g: S \to U$ is also onto.

An immediate consequence of combining the Remark and Lemma 1.3.2 is to obtain

Lemma 1.3.3. If $g: S \to T$ and $f: T \to U$ are both bijec-tions, then $f \circ g: S \to U$ is also a bijection.

If f is a 1-1 correspondence of S onto T, then the "object" $f^{-1}: T \to S$ defined earlier can easily be shown to be a 1-1 mapping of T onto S. In this case it is called the *inverse* of f. In this situation we have

Lemma 1.3.4. If $f: S \to T$ is a bijection, then $f \circ f^{-1} = i_T$ and $f^{-1} \circ f = i_S$, where i_S and i_T are the identity mappings of S and T, respectively.

Proof. We verify one of these. If $t \in T$, then $(f \circ f^{-1})(t) = f(f^{-1}(t))$. But what is $f^{-1}(t)$? By definition, $f^{-1}(t)$ is that element $s_0 \in S$ such that $t = f(s_0)$. So $f(f^{-1}(t)) = f(s_0) = t$. In other words, $(f \circ f^{-1})(t) = t$ for every $t \in T$; hence $f \circ f^{-1} = i_T$, the identity mapping on T. $\square$

We leave the last result of this section for the reader to prove.

Lemma 1.3.5. If $f: S \to T$ and i_T is the identity mapping of T into itself and i_S is that of S onto itself, then $i_T \circ f = f$ and $f \circ i_S = f$.

Problems

Easier Problems

1. For the given S, T determine if a mapping $f: S \to T$ is clearly and unambiguously defined; if not, say why not.
   (a) S = set of all women, T = set of all men, $f(s)$ = husband of s.
   (b) S = set of positive integers, $T = S$, $f(s) = s - 1$.
   (c) S = set of positive integers, T = set of nonnegative integers, $f(s) = s - 1$.
   (d) S = set of nonnegative integers, $T = S$, $f(s) = s - 1$.
   (e) S = set of all integers, $T = S$, $f(s) = s - 1$.
   (f) S = set of all real numbers, $T = S$, $f(s) = \sqrt{s}$.
   (g) S = set of all positive real numbers, $T = S$, $f(s) = \sqrt{s}$.

2. In those parts of Problem 1 where f does define a function, determine if it is 1-1, onto, or both.

*3. If f is a 1-1 mapping of S onto T, prove that f^{-1} is a 1-1 mapping of T onto S.

*4. If f is a 1-1 mapping of S onto T, prove that $f^{-1} \circ f = i_S$.

5. Give a proof of the Remark after Lemma 1.3.2.

*6. If $f: S \to T$ is onto and $g: T \to U$ and $h: T \to U$ are such that $g \circ f = h \circ f$, prove that $g = h$.

*7. If $g: S \to T$, $h: S \to T$, and if $f: T \to U$ is 1-1, show that if $f \circ g = f \circ h$, then $g = h$.

8. Let S be the set of all integers and $T = \{1, -1\}$; $f: S \to T$ is defined by $f(s) = 1$ if s is even, $f(s) = -1$ if s is odd.
   (a) Does this define a function from S to T?
   (b) Show that $f(s_1 + s_2) = f(s_1)f(s_2)$. What does this say about the integers?
   (c) Is $f(s_1 s_2) = f(s_1)f(s_2)$ also true?

9. Let S be the set of all real numbers. Define $f: S \rightarrow S$ by $f(s) = s^2$, and $g: S \rightarrow S$ by $g(s) = s + 1$.
   (a) Find $f \circ g$.
   (b) Find $g \circ f$.
   (c) Is $f \circ g = g \circ f$?

10. Let S be the set of all real numbers and for $a, b \in S$, where $a \neq 0$; define $f_{a,b}(s) = as + b$.
   (a) Show that $f_{a,b} \circ f_{c,d} = f_{u,v}$ for some real u, v. Give explicit values for u, v in terms of a, b, c, and d.
   (b) Is $f_{a,b} \circ f_{c,d} = f_{c,d} \circ f_{a,b}$ always?
   (c) Find all $f_{a,b}$ such that $f_{a,b} \circ f_{1,1} = f_{1,1} \circ f_{a,b}$.
   (d) Show that $f_{a,b}^{-1}$ exists and find its form.

11. Let S be the set of all positive integers. Define $f: S \rightarrow S$ by $f(1) = 2$, $f(2) = 3$, $f(3) = 1$, and $f(s) = s$ for any other $s \in S$. Show that $f \circ f \circ f = i_S$. What is f^{-1} in this case?

Middle-Level Problems

12. Let S be the set of all nonnegative rational numbers, that is, $S = \{ m/n \mid m, n \text{ nonnegative integers}, n \neq 0 \}$, and let T be the set of all integers.
   (a) Does $f: S \rightarrow T$ defined by $f(m/n) = 2^m 3^n$ define a legitimate function from S to T?
   (b) If not, how could you modify the definition of f so as to get a legitimate function?

13. Let S be the set of all positive integers of the form $2^m 3^n$, where $m > 0$, $n > 0$, and let T be the set of all rational numbers. Define $f: S \rightarrow T$ by $f(2^m 3^n) = m/n$. Prove that f defines a function from S to T. (On what properties of the integers does this depend?)

14. Let $f: S \rightarrow S$, where S is the set of all integers, be defined by $f(s) = as + b$, where a, b are integers. Find the necessary and sufficient conditions on a, b in order that $f \circ f = i_S$.

15. Find all f of the form given in Problem 14 such that $f \circ f \circ f = i_S$.

16. If f is a 1-1 mapping of S onto itself, show that $(f^{-1})^{-1} = f$.

17. If S is a finite set having $m > 0$ elements, how many mappings are there of S into itself?

18. In Problem 17, how many 1-1 mappings are there of S into itself?

19. Let S be the set of all real numbers, and define $f: S \rightarrow S$ by $f(s) = s^2 + as + b$, where a, b are fixed real numbers. Prove that for no values of a, b can f be onto or 1-1.

20. Let S be the set of all positive real numbers. For a, b, c, d real numbers and c, d positive, is it ever possible that the "mapping" $f: S \rightarrow S$ defined by $f(s) = (as + b)/(cs + d)$ satisfies $f \circ f = i_S$? Find all a, b, c, d that do the trick. Is f a mapping of S into itself?

21. Let S be the set of all rational numbers and let $f_{a,b} : S \rightarrow S$ be defined by $f_{a,b}(s) = as + b$, where $a \neq 0$, b are rational numbers. Find all $f_{c,d}$ of this form satisfying $f_{c,d} \circ f_{a,b} = f_{a,b} \circ f_{c,d}$ for *every* $f_{a,b}$.

22. Let S be the set of all integers and a, b, c rational numbers. Define $f: S \rightarrow S$ by $f(s) = as^2 + bs + c$. Find necessary and sufficient conditions on a, b, c, so that f defines a mapping on S [**Note:** a, b, c need not be integers; for example, $f(s) = \frac{1}{2}s(s + 1) = \frac{1}{2}s^2 + \frac{1}{2}s$ *does* always give us an integer for integral s.]

Harder Problems

23. Let S be the set of all integers of the form $2^m 3^n$, $m \geq 0$, $n \geq 0$, and let T be the set of all positive integers. Show that there is a 1-1 correspondence of S onto T.

24. Prove that there is a 1-1 correspondence of the set of all positive integers onto the set of all positive rational numbers.

25. Let S be the set of all real numbers and T the set of all positive reals. Find a 1-1 mapping f of S onto T such that $f(s_1 + s_2) = f(s_1)f(s_2)$ for all $s_1, s_2 \in S$.

26. For the f in Problem 25, find f^{-1} explicitly.

27. If f, g are mappings of S into S and $f \circ g$ is a *constant* function, then

    (a) What can you say about f if g is onto?
    (b) What can you say about g if f is 1-1?

28. If S is a finite set and f is a mapping of S *onto* itself, show that f must be 1-1.

29. If S is a finite set and f is a 1-1 mapping of S into itself, show that f must be onto.

30. If S is a finite set and f is a 1-1 mapping of S, show that for some integer $n > 0$,

$$\underbrace{f \circ f \circ f \circ \cdots \circ f}_{n \text{ times}} = i_S.$$

31. If S has m elements in Problem 30, find an $n > 0$ (in terms of m) that works simultaneously for all 1-1 mappings of S into itself.

4. $A(S)$ (THE SET OF 1-1 MAPPINGS OF S ONTO ITSELF)

We focus our attention in this section on particularly nice mappings of a nonempty set, S, into itself. Namely, we shall consider the set, $A(S)$, of all 1-1 mappings of S onto itself. Although most of the concern in the book will be in the case in which S is a finite set, we do not restrict ourselves to that situation here.

When S has a finite number of elements, say n, then $A(S)$ has a special name. It is called the *symmetric group of degree n* and is often denoted by S_n. Its elements are called *permutations* of S. If we are interested in the structure of S_n, it really does not matter much what our underlying set S is. So, you can think of S as being the set $(1, \ldots, n)$. Chapter 3 will be devoted to a study, in some depth, of S_n. In the investigation of finite groups, S_n plays a central role.

There are many properties of the set $A(S)$ on which we could concentrate. We have chosen to develop those aspects here which will motivate the notion of a group and which will give the reader some experience, and feeling for, working in a group-theoretic framework. Groups will be discussed in Chapter 2.

We begin with a result that is really a compendium of some of the results obtained in Section 3.

Lemma 1.4.1. $A(S)$ satisfies the following:
  (a) $f, g \in A(S)$ implies that $f \circ g \in A(S)$.
  (b) $f, g, h \in A(S)$ implies that $(f \circ g) \circ h = f \circ (g \circ h)$.

(c) There exists an element—the identity mapping i—such that $f \circ i = i \circ f = f$ for every $f \in A(S)$.

(d) Given $f \in A(S)$, there exists a $g \in A(S)$ $(g = f^{-1})$ such that $f \circ g = g \circ f = i$.

Proof. All these things were done in Section 3, either in the text material or in the problems. We leave it to the reader to find the relevant part of Section 3 that will verify each of the statements (a) through (d). ☐

We should now like to know how many elements there are in $A(S)$ when S is a finite set having n elements. To do so, we first carry out a slight digression.

Suppose that you can do a certain thing in r different ways and a second independent thing in s different ways. In how many distinct ways can you do both things together? The best way of finding out is to picture this in a concrete context. Suppose that there are r highways running from Chicago to Detroit and s highways running from Detroit to Ann Arbor. In how many ways can you go first to Detroit, then to Ann Arbor? Clearly, for every road you take from Chicago to Detroit you have s ways of continuing on to Ann Arbor. You can start your trip from Chicago in r distinct ways, hence you can complete it in

$$\underbrace{s + s + s + \cdots + s}_{r \text{ times}} = rs$$

different ways.

It is fairly clear that we can extend this from doing two independent things to doing m independent ones, for any integer $m > 2$. If we can do the first thing in r_1 distinct ways, the second in r_2 ways, ..., the mth in r_m distinct ways, then we can do all these together in $r_1 r_2 \ldots r_m$ different ways.

Let's recall something many of us have already seen:

Definition. If $n \geq 1$ is a positive integer, then $n!$ (read "*n factorial*") is defined by $n! = 1 \cdot 2 \cdot 3 \cdots n$.

Lemma 1.4.2. If S has n elements, then $A(S)$ has $n!$ elements.

Proof. Let $f \in A(S)$, where $S = \{x_1, x_2, \ldots, x_n\}$. How many choices does f have as a place to send x_1? Clearly n, for we can send x_1 under f to *any* element of S. But now f is *not* free to send x_2 anywhere, for since f is 1-1, we must have $f(x_1) \neq f(x_2)$. So we can send x_2 anywhere except onto $f(x_1)$. Hence f can send x_2 into $n - 1$ different images. Continuing this way, we see that f can send x_i into $n - (i - 1)$ different images. Hence the number of such f's is $n(n - 1)(n - 2) \cdots 1 = n!$. $\square$

EXAMPLE

The number $n!$ gets very large quickly. To be able to see the picture in its entirety, we look at the special case $n = 3$, where $n!$ is still quite small.

Consider $A(S) = S_3$, where S consists of the three elements x_1, x_2, x_3. We list all the elements of S_3, writing out each mapping explicitly by what it does to each of x_1, x_2, x_3.

1. $i : x_1 \rightarrow x_1, x_2 \rightarrow x_2, x_3 \rightarrow x_3$.
2. $f : x_1 \rightarrow x_2, x_2 \rightarrow x_3, x_3 \rightarrow x_1$.
3. $g : x_1 \rightarrow x_2, x_2 \rightarrow x_1, x_3 \rightarrow x_3$.
4. $g \circ f : x_1 \rightarrow x_1, x_2 \rightarrow x_3, x_3 \rightarrow x_2$. (Verify!)
5. $f \circ g : x_1 \rightarrow x_3, x_2 \rightarrow x_2, x_3 \rightarrow x_1$. (Verify!)
6. $f \circ f : x_1 \rightarrow x_3, x_2 \rightarrow x_1, x_3 \rightarrow x_2$. (Verify!)

Since we have listed here six distinct elements of S_3, and S_3 has only six elements, we have a complete list of all the elements of S_3. What does this list tell us? To begin with, we note that $f \circ g \neq g \circ f$, so one familiar rule of the kind of arithmetic we have been used to is violated. Since $g \in S_3$ and $g \in S_3$, we must have $g \circ g$ also in S_3. What is it? If we calculate $g \circ g$, we easily get $g \circ g = i$. Similarly, we get

$$(f \circ g) \circ (f \circ g) = i = (g \circ f) \circ (g \circ f).$$

Note also that $f \circ (f \circ f) = i$, hence $f^{-1} = f \circ f$. Finally, we leave it to the reader to show that $g \circ f = f^{-1} \circ g$.

It is a little cumbersome to write this product in $A(S)$ using the $\circ$. *From now on we shall drop it and write $f \circ g$ merely as fg.* Also, we shall start using the shorthand of exponents, to avoid expressions like $f \circ f \circ f \circ \cdots \circ f$. We define, for $f \in A(S)$, $f^0 = i$, $f^2 = f \circ f = ff$, and so on. For negative exponents $-n$ we define f^{-n} by $f^{-n} = (f^{-1})^n$, where n is a positive integer. The usual rules of exponents prevail, namely $f^r f^s = f^{r+s}$ and $(f^r)^s = f^{rs}$. We leave these as exercises—somewhat tedious ones at that—for the reader.

EXAMPLE

Do not jump to conclusions that all familiar properties of exponents go over. For instance, in the example of the $f, g \in S_3$ defined above we claim that $(fg)^2 \neq f^2 g^2$. To see this, we note that

$$fg: x_1 \rightarrow x_3, x_2 \rightarrow x_2, x_3 \rightarrow x_1,$$

so that $(fg)^2 : x_1 \rightarrow x_1, x_2 \rightarrow x_2, x_3 \rightarrow x_3$, that is, $(fg)^2 = i$. On the other hand, $f^2 \neq i$ and $g^2 = i$, hence $f^2 g^2 = f^2 \neq i$, whence $(fg)^2 \neq f^2 g^2$ in this case.

However, some other familiar properties do go over. For instance, if f, g, h are in $A(S)$ and $fg = fh$, then $g = h$. Why? Because, from $fg = fh$ we have $f^{-1}(fg) = f^{-1}(fh)$; therefore, $g = ig = (f^{-1}f)g = f^{-1}(fg) = f^{-1}(fh) = (f^{-1}f)h = ih = h$. Similarly, $gf = hf$ implies that $g = h$. So we can cancel an element in such an equation provided that we *do not change sides*. In S_3 our f, g there satisfy $gf = f^{-1}g$, but since $f \neq f^{-1}$ we *cannot cancel* the g here.

Problems

Recall that fg stands for $f \circ g$ and, also, what f^m means. S, without subscripts, will be a nonempty set.

Easier Problems

1. If $s_1 \neq s_2$ are in S, show that there is an $f \in A(S)$ such that $f(s_1) = s_2$.

2. If $s_1 \in S$, let $H = \{ f \in A(S) | f(s_1) = s_1 \}$. Show that:
   (a) $i \in H$.
   (b) If $f, g \in H$, then $fg \in H$.
   (c) If $f \in H$, then $f^{-1} \in H$.

3. Suppose that $s_1 \neq s_2$ are in S and $f(s_1) = s_2$, where $f \in A(S)$. Then if H is as in Problem 2 and $K = \{g \in A(S) | g(s_2) = s_2\}$, show that:
   (a) If $g \in K$, then $f^{-1}gf \in H$.
   (b) If $h \in H$, then there is some $g \in K$ such that $h = f^{-1}gf$.

4. If $f, g, h \in A(S)$, show that $(f^{-1}gf)(f^{-1}hf) = f^{-1}(gh)f$. What can you say about $(f^{-1}gf)^n$?

5. If $f, g \in A(S)$ and $fg = gf$, show that:
   (a) $(fg)^2 = f^2g^2$.
   (b) $(fg)^{-1} = f^{-1}g^{-1}$.

6. Push the result of Problem 5, for the same f and g, to show that $(fg)^m = f^mg^m$ for all integers m.

*7. Verify the rules of exponents, namely $f^rf^s = f^{r+s}$ and $(f^r)^s = f^{rs}$ for $f \in A(S)$ and positive integers r, s.

8. If $f, g \in A(S)$ and $(fg)^2 = f^2g^2$, prove that $fg = gf$.

9. If $S = \{x_1, x_2, x_3, x_4\}$, let $f, g \in S_4$ be defined by

$$f: x_1 \to x_2, x_2 \to x_3, x_3 \to x_4, x_4 \to x_1,$$

and

$$g: x_1 \to x_2, x_2 \to x_1, x_3 \to x_3, x_4 \to x_4.$$

Calculate:
(a) f^2, f^3, f^4.
(b) g^2, g^3.
(c) fg.
(d) gf.
(e) $(fg)^3, (gf)^3$.
(f) Are fg and gf equal?

10. If $f \in S_3$, show that $f^6 = i$.

11. Can you find a positive integer m such that $f^m = i$ for *all* $f \in S_4$?

Middle-Level Problems

*12. If $f \in S_n$, show that there is some positive integer k, depending on f, such that $f^k = i$. (**Hint:** Consider the positive powers of f.)

*13. Show that there is a positive integer t such that $f^t = i$ for *all* $f \in S_n$.

14. If $m < n$, show that there is a 1-1 mapping $F: S_m \to S_n$ such that $F(fg) = F(f)F(g)$ for all $f, g \in S_m$.

15. If S has three or more elements, show that we can find $f, g \in A(S)$ such that $fg \neq gf$.

16. Let S be an infinite set and let $M \subset A(S)$ be the set of all elements $f \in A(S)$ such that $f(s) \neq s$ for at most a finite number of $s \in S$. Prove that:
    (a) $f, g \in M$ implies that $fg \in M$.
    (b) $f \in M$ implies that $f^{-1} \in M$.

17. For the situation in Problem 16, show, if $f \in A(S)$, that $f^{-1}Mf = \{f^{-1}gf | g \in M\}$ must equal M.

18. Let $S \supset T$ and consider the subset $U(T) = \{f \in A(S) | f(t) \in T$ for every $t \in T\}$. Show that:
    (a) $i \in U(T)$.
    (b) $f, g \in U(T)$ implies that $fg \in U(T)$.

19. If the S in Problem 18 has n elements and T has m elements, how many elements are there in $U(T)$? Show that there is a mapping $F: U(T) \to S_m$ such that $F(fg) = F(f)F(g)$ for $f, g \in U(T)$ and F is onto S_m.

20. If $m < n$, can F in Problem 19 ever be 1-1? If so, when?

21. In S_n show that the mapping f defined by
$$f: x_1 \to x_2, x_2 \to x_3, x_3 \to x_4, \ldots, x_{n-1} \to x_n, x_n \to x_1$$
[i.e., $f(x_i) = x_{i+1}$ if $i < n$, $f(x_n) = x_1$] can be written as $f = g_1 g_2 \cdots g_{n-1}$ where each $g_i \in S_n$ *interchanges exactly two elements of* $S = \{x_1, \ldots, x_n\}$, leaving the other elements fixed in S.

Harder Problems

22. If $f \in S_n$, show that $f = h_1 h_2 \cdots h_m$ for some $h_j \in S_n$ such that $h_j^2 = i$.

*23. Call an element in S_n a *transposition* if it interchanges two elements, leaving the others fixed. Show that any element in S_n is a product of transpositions. (This sharpens the result of Problem 22.)

24. If n is at least 3, show that for some f in S_n, f cannot be expressed in the form $f = g^3$ for any g in S_n.

25. If $f \in S_n$ is such that $f \neq i$ but $f^3 = i$, show that we can number the elements of S in such a way that $f(x_1) = x_2$, $f(x_2) = x_3$, $f(x_3) = x_1$, $f(x_4) = x_5$, $f(x_5) = x_6$, $f(x_6) = x_4, \ldots, f(x_{3k+1}) = x_{3k+2}$, $f(x_{3k+2}) = x_{3k+3}$, $f(x_{3k+3}) = x_{3k+1}$, for some k, and for all the other $x_t \in S$ $f(x_t) = x_t$.

26. View a fixed shuffle of a deck of 52 cards as a 1-1 mapping of the deck onto itself. Show that repeating this fixed shuffle a finite (positive) number of times will bring the deck back to its original order.

*27. If $f \in A(S)$, call, for $s \in S$, the *orbit* of s (relative to f) the set $O(s) = \{ f^j(s) | \text{all integers } j \}$. Show that if $s, t \in S$, then either $O(s) \cap O(t) = \varnothing$ or $O(s) = O(t)$.

28. If $S = \{ x_1, x_2, \ldots, x_{12} \}$ and $f \in S_{12}$ is defined by $f(x_i) = x_{i+1}$ if $i = 1, 2, \ldots, 11$ and $f(x_{12}) = x_1$, find the orbits of all the elements of S (relative to f).

29. If $f \in A(S)$ satisfies $f^3 = i$, show that the orbit of any element of S has one or three elements.

*30. Recall that a positive integer $p > 1$ is called a *prime number* if p cannot be factored as a product of smaller positive integers. If $f \in A(S)$ satisfies $f^p = i$, what can you say about the size of the orbits of the elements of S relative to f? What property of the prime numbers are you using to get your answer?

31. Prove that if S has more than two elements, then the only elements f_0 in $A(S)$ such that $f_0 f = f f_0$ for all $f \in A(S)$ must satisfy $f_0 = i$.

*32. We say that $g \in A(S)$ *commutes* with $f \in A(S)$ if $fg = gf$. Find all elements in $A(S)$ that commute with $f : S \to S$ defined by $f(x_1) = x_2$, $f(x_2) = x_1$, and $f(s) = s$ if $s \neq x_1, x_2$.

33. In S_n show that the only elements commuting with f defined by $f(x_i) = x_{i+1}$ if $i < n$, $f(x_n) = x_1$, are the powers of f, namely $i = f^0, f, f^2, \ldots, f^{n-1}$.

34. Let $C(f) = \{ g \in A(S) | fg = gf \}$. Prove that:
    (a) $g, h \in C(f)$ implies that $gh \in C(f)$.
    (b) $g \in C(f)$ implies that $g^{-1} \in C(f)$.
    (c) $C(f)$ is not empty.

5. THE INTEGERS

The mathematical set most familiar to everybody is that of the positive
integers $1, 2, \ldots$, which we shall often call N. Equally familiar is the set,
$\mathbb{Z}$, of all integers—positive, negative, and zero. Because of this acquaintance with $\mathbb{Z}$, we shall give here a rather sketchy survey of the properties
of $\mathbb{Z}$ that we shall use often in the ensuing material. Most of these
properties are well known to all of us; a few are less well known.

The basic assumption we make about the set of integers is the

Well-Ordering Principle.   Any nonempty set of nonnegative integers has a smallest member.

More formally, what this principle states is that given a nonempty
set V of nonnegative integers, there is an element $v_0 \in V$ such that $v_0 \le$
v for every $v \in V$. This principle will serve as the foundation for the
development about the integers that we are about to make.

The first application we make of it is to show something we all know
and have taken for granted, namely that we can divide one integer by
another to get a remainder that is smaller. This is known as *Euclid's
Algorithm*. We give it a more formal statement and a proof based on
well-ordering.

Theorem 1.5.1 (Euclid's Algorithm).   If m and n are integers
with $n > 0$, then there exist integers q and r, with $0 \le r < n$,
such that $m = qn + r$.

Proof.   Let W be the set $m - tn$, where t runs through all the integers
(i.e., $W = \{ m - tn \mid t \in Z \}$). We claim that W contains some nonnegative integers, for if t is large enough and negative, then $m - tn > 0$. Let
$V = \{ v \in W \mid v \ge 0 \}$; by the well-ordering principle V has a smallest
element, r. Since $r \in V$, $r \ge 0$, and $r = m - qn$ for some q (for that is
the form of all elements in $W \supset V$). We claim that $r < n$. If not,
$r = m - qn \ge n$, hence $m - (q + 1)n \ge 0$. But this puts $m - (q + 1)n$
in V, yet $m - (q + 1)n < r$, contradicting the minimal nature of r in V.
With this, Euclid's Algorithm is proved. $\square$

Euclid's Algorithm will have a host of consequences for us, especially about the nótion of divisibility. Since we are speaking about the integers, *be it understood that all letters used in this section will be integers.* This will save a lot of repetition of certain phrases.

> **Definition.**   Given integers $m \neq 0$ and n we say that m *divides n*, written as $m|n$, if $n = cm$ for some integer c.

Thus, for instance, $2|14$, $(-7)|14$, $4|(-16)$. If $m|n$, we call m a *divisor* or *factor* of n, and n a *multiple* of m. To indicate that m is not a divisor of n, we write $m \nmid n$; so, for instance, $3 \nmid 5$.

The basic elementary properties of divisibility are laid out in

> **Lemma 1.5.2**   The following are true:
>   (a) $1|n$ for all n.
>   (b) If $m \neq 0$, then $m|0$.
>   (c) If $m|n$ and $n|q$, then $m|q$.
>   (d) If $m|n$ and $m|q$, then $m|(un + vq)$ for all u, v.
>   (e) If $m|1$, then $m = 1$ or $m = -1$.
>   (f) If $m|n$ and $n|m$, then $m = \pm n$.

Proof.   The proofs of all these parts are easy, following immediately from the definition of $m|n$. We leave all but Part (d) as exercises but prove Part (d) here to give the flavor of how such proofs go.

So suppose that $m|n$ and $m|q$.   Then $n = cm$ and $q = dm$ for some c and d $un + vq = u(cm) + v(dm) = (uc + vd)m$.   Thus, from the definition, $m|(un + vq)$. □

Having the concept of a divisor of an integer, we now want to introduce that of the *greatest common divisor* of two (or more) integers. Simply enough, this should be the largest possible integer that is a divisor of both integers in question. However, we want to avoid using the size of an integer—for reasons that may become clear much later when we talk about rings. So we make the definition in what may seem as a strange way.

Definition. Given a, b (not both 0), then their *greatest common divisor* c is defined by:
  (a) $c > 0$.
  (b) $c|a$ and $c|b$.
  (c) If $d|a$ and $d|b$, then $d|c$.
We write this c as $c = (a, b)$.

In other words, the greatest common divisor of a and b is the positive number c which divides a and b and is divisible by every d which divides a and b.

Defining something does not guarantee its existence. So it is incumbent on us to prove that (a, b) exists, and is, in fact, unique. The proof actually shows more, namely that (a, b) is a nice combination of a and b. This combination is not unique; for instance,

$$(24, 9) = 3 = 3 \cdot 9 + (-1)24 = (-5)9 + 2 \cdot 24.$$

Theorem 1.5.3. If a, b are not both 0, then their *greatest common divisor* $c = (a, b)$ exists, is unique, and, moreover, $c = m_0 a + n_0 b$ for some suitable m_0 and n_0.

Proof. Since not both a and b are 0, the set $A = \{ma + nb | m, n \in Z\}$ has nonzero elements. If $x \in A$ and $x < 0$, then $-x$ is also in A and $-x > 0$, for if $x = m_1 a + n_1 b$, then $-x = (-m_1)a + (-n_1)b$, so is in A. Thus A has positive elements; hence, by the well-ordering principle there is a smallest positive element, c, in A. Since $c \in A$, by the form of the elements of A we know that $c = m_0 a + n_0 b$ for some m_0, n_0.

We claim that c is our required greatest common divisor. First note that if $d|a$ and $d|b$, then $d|(m_0 a + n_0 b)$ by Part (d) of Lemma 1.5.2, that is, $d|c$. So, to verify that c is our desired element, we need only show that $c|a$ and $c|b$.

By Euclid's Algorithm, $a = qc + r$, where $0 \le r < c$, that is, $a = q(m_0 a + n_0 b) + r$. Therefore, $r = -qn_0 b + (1 - qm_0)a$. So r is in A. But $r < c$ and is in A, so by the choice of c, r *cannot be* positive. Hence $r = 0$; in other words, $a = qc$ and so $c|a$. Similarly, $c|b$.

For the uniqueness of c, if $t > 0$ also satisfied $t|a, t|b$ and $d|t$ for all d such that $d|a$ and $d|b$, we would have $t|c$ and $c|t$. By Part (f) of Lemma 1.5.2 we get that $t = c$ (since both are positive). $\square$

Let's look at an explicit example, namely $a = 24$, $b = 9$. By direct examination we know that $(24, 9) = 3$; note that $3 = 3 \cdot 9 + (-1)24$. What is $(-24, 9)$?

How is this done for positive numbers a and b which may be quite large? If $b > a$, interchange a and b so that $a > b > 0$. The we can find (a, b) by

1. observing that $(a, b) = (b, r)$ where $a = qb + r$ with $0 \le r < b$ (Why?);

2. finding (b, r), which now is easier since one of the numbers is smaller than before.

So, for example, we have

$(100, \ 28) = (28, \ 16)$	since $100 = 3\ (28) + 16$
$(\ 28, \ 16) = (16, \ 12)$	since $\ 28 = 1\ (16) + 12$
$(\ 16, \ 12) = (16, \ \ 4)$	since $\ 16 = 1\ (12) + \ \ 4$

This gives us
$$(100, 28 = (16, \ \ 4) = 4.$$

It is possible to find the actual values of m_0 and n_0 such that
$$4 = m_0\ 100 + n_0\ 28$$

by going backwards through the calculations made to find 4:

Since $\ 16 = 1\ (12) + \ \ 4,$	$4 = \ \ 16 + (\ -1\ )\ 12$
Since $\ 28 = 1\ (16) + 12,$	$12 = \ \ 28 + (\ -1\ )\ 16$
Since $100 = 3\ (28) + 16,$	$16 = 100 + (\ -3\ )\ 28$

But then

$$4 = \qquad 16 + (-1)\ 12 = \qquad 16 + (\ -1\ )\ (\ 28\ + (-1)\ 16)$$
$$= (-1)\ \ 28 + (\ \ 2)\ 16 = (-1)\ 28 + (\ \ 2)\ (100$$
$$= (\ \ 2)\ 100 + (-7)\ 28$$

so that $m_0 = 2$ and $n_0 = -7$

This method of using steps (1) and (2) over and over again shows how to use Euclid's Algorithm can be used to compute (a, b) for any positive integers a and b.

We shall include some exercises at the end of this section on other properties of (a, b).

We come to the very important

Definition. We say that a and b are *relatively prime* if $(a, b) = 1$.

So the integers a and b are relatively prime if they have no nontrivial common factor. An immediate corollary to Theorem 1.5.3 is

Theorem 1.5.4. The integers a and b are relatively prime if and only if $1 = ma + nb$ for suitable integers m and n.

Theorem 1.5.4 has an immediate consequence

Theorem 1.5.5. If a and b are relatively prime and $a|bc$, then $a|c$.

Proof. By Theorem 1.5.4, $ma + nb = 1$ for some m and n, hence $(ma + nb)c = c$, that is, $mac + nbc = c$. By assumption, $a|bc$ and by observation $a|mac$, hence $a|(mac + nbc)$ and so $a|c$. $\square$

Corollary. If b and c are both relatively prime to a, then bc is also relatively prime to a.

Proof. We pick up the proof of Theorem 1.5.5 at $mac + nbc = c$. If $d = (a, bc)$, then $d|a$ and $d|bc$, hence $d|(mac + nbc) = c$. Since $d|a$ and $d|c$ and $(a, c) = 1$, we get that $d = 1$. Since $1 = d = (a, bc)$, we have that bc is relatively prime to a. $\square$

We now single out an ultra-important class of positive integers, which we met before in Problem 30, Section 4.

Definition. The integer $p > 1$ is a *prime number*, or *prime*, if for any integer a either $p|a$ or p is relatively prime to a.

This definition coincides with the usual one, namely that we cannot factor p nontrivially. For if p is a prime as defined above and $p = ab$ where $1 \le a < b$, then $(a, p) = a$ (Why?) and p does not divide a since $p > a$. It follows that $a = 1$, so $p = b$. On the other hand, if p is a prime in the sense that it cannot be factored nontrivially, and if a is an integer not relatively prime to p, then (a, b) is not 1 and it divides a and p. But then (a, b) equals p, by our hypothesis, so p divides a.

Another result coming out of Theorem 1.5.5 is

Theorem 1.5.6. If p is a prime and $p|(a_1a_2 \cdots a_n)$, then $p|a_i$ for some i with $1 \le i \le n$.

Proof. If $p|a_1$, there is nothing to prove. Suppose that $p \nmid a_1$; then p and a_1 are relatively prime. But $p|a_1(a_2 \cdots a_n)$ hence by Theorem 1.5.5, $p|a_2 \cdots a_n$. Repeat the argument just given on a_2, and continue. $\square$

The primes play a very special role in the set of integers larger than 1 in that every integer $n > 1$ is either a prime or is the product of primes. We shall show this in the next theorem. In the theorem after the next we shall show that there is a uniqueness about the way $n > 1$ factors into prime factors. The proofs of both these results lean heavily on the well-ordering principle.

Theorem 1.5.7. If $n > 1$, then either n is a prime or n is the product of primes.

Proof. Suppose that the theorem is false. Then there must be an integer $m > 1$ for which the theorem fails. Therefore, the set M for which the theorem fails is nonempty, so, by the well-ordering principle, M has a least element m. Clearly, since $m \in M$, m cannot be a prime, thus $m = ab$, where $1 < a < m$ and $1 < b < m$. Because $a < m$ and $b < m$ and m is the least element in M, we cannot have $a \in M$ or $b \in M$. Since $a \notin M$, $b \notin M$, by the definition of M the theorem must be true for both a and b. Thus a and b are primes or the product of primes; from $m = ab$ we get that m is a product of primes. This puts m outside of M, contradicting that $m \in M$. This proves the theorem. $\square$

We asserted above that there is a certain uniqueness about the decomposition of an integer into primes. We make this precise now. To avoid trivialities of the kind $6 = 2 \cdot 3 = 3 \cdot 2$ (so, in a sense, 6 has two factorizations into the primes 2 and 3), we shall state the theorem in a particular way.

Theorem 1.5.8. Given $n > 1$, then there is one and only one way to write n in the form $n = p_1^{a_1} p_2^{a_2} \cdots p_k^{a_k}$, where $p_1 < p_2 < \cdots p_k$ are primes and the exponents $a_1, a_2, \ldots, a_k$ are all positive.

Proof. We start as we did above by assuming that the theorem is false, so there is a least integer $m > 1$ for which it is false. This m *must have two distinct factorizations* as $m = p_1^{a_1} p_2^{a_2} \cdots p_k^{a_k} = q_1^{b_1} q_2^{b_2} \cdots q_\ell^{b_\ell}$, where $p_1 < p_2 < \cdots < p_k$, $q_1 < q_2 \cdots q_\ell$ are primes and where the exponents $a_1, \ldots, a_k$ and $b_1, \ldots, b_\ell$ are all positive. Since $p_1 | p_1^{a_1} \cdots p_k^{a_k} = q_1^{b_1} \cdots q_\ell^{b_\ell}$, by Theorem 1.5.6 $p_1 | q_i^{b_i}$ for some i; hence, again by Theorem 1.5.6, $p_1 | q_i$, hence $p_1 = q_i$. By the same token $q_1 = p_j$ for some j; thus $p_1 \leq p_j = q_1 \leq q_i = p_1$. This gives us that $p_1 = q_1$. Now since $m/p_1 < m$, m/p_1 has the unique factorization property. But $m/p_1 = p_1^{a_1-1} p_2^{a_2} \cdots p_k^{a_k} = p_1^{b_1-1} q_2^{b_2} \cdots q_\ell^{b_\ell}$ and since m/p_1 can be factored in one and only one way in this form, we easily get $k = \ell$, $p_2 = q_2, \ldots, p_k = q_k$, $a_1 - 1 = b_1 - 1$, $a_2 = b_2, \ldots, a_k = b_k$. So we see that the primes and their exponents arising for the factorization of m are unique. This contradicts the lack of such uniqueness for m, and so proves the theorem. $\square$

What these last two theorems tell us is that we can build up the integers from the primes in a very precise and well-defined manner. One would expect from this that there should be many—that is, an infinity—of primes. This old result goes back to Euclid; in fact, the argument we shall give is due to Euclid.

Theorem 1.5.9. There is an infinite number of primes.

Proof. If the result were false, we could enumerate *all* the primes in $p_1, p_2, \ldots, p_k$. Consider the integer $q = 1 + p_1 p_2 \cdots p_k$. Since $q > p_i$ for every $i = 1, 2, \ldots, k$, q cannot be a prime. Since $p_i \nmid q$, for we get a remainder of 1 on dividing q by p_i, q is not divisible by any of

$p_1, \ldots, p_k$. So q is not a prime nor is it divisible by any prime. This violates Theorem 1.5.7, thereby proving the theorem. $\square$

Results much sharper than Theorem 1.5.9 exist about how many primes there are up to a given point. The famous prime number theorem states that for large n the number of primes less than or equal to n is "more or less" $n/\log_e n$, where this "more or less" is precisely described. There are many open questions about the prime numbers.

Problems

Easier Problems

1. Find (a, b) and express (a, b) as $ma + nb$ for:
   (a) $(116, -84)$.
   (b) $(85, 65)$.
   (c) $(72, 26)$.
   (d) $(72, 25)$.

2. Prove all the parts of Lemma 1.5.2.

3. Show that $(ma, mb) = m(a, b)$ if $m > 0$.

4. Show that if $a|m$ and $b|m$ and $(a, b) = 1$, then $(ab)|m$.

5. Factor the following into primes.
   (a) 36.
   (b) 120.
   (c) 720.
   (d) 5040.

6. If $m = p_1^{a_1} \cdots p_k^{a_k}$ and $n = p_1^{b_1} \cdots p_k^{b_k}$, where $p_1, \ldots, p_k$ are distinct primes and $a_1, \ldots, a_k$ are nonnegative and $b_1, \ldots, b_k$ are nonnegative, express (m, n) as $p_1^{c_1} \cdots p_k^{c_k}$ by describing the c's in terms of the a's and b's.

*7. Define the *least common multiple* of positive integers m and n to be the smallest positive integer v such that both $m|v$ and $n|v$.
   (a) Show that $v = mn/(m, n)$.
   (b) In terms of the factorization of m and n given in Problem 6, what is v?

8. Find the least common multiple of the pairs given in Problem 1.

9. If $m, n > 0$ are two integers, show that we can find integers u, v with $-n/2 \le v \le n/2$ such that $m = un + v$.

10. To check that a given integer $n > 1$ is a prime, prove that it is enough to show that n is not divisible by any prime p with $p \le \sqrt{n}$.

11. Check if the following are prime.
   (a) 301.
   (b) 1001.
   (c) 473.

12. Starting with $2, 3, 5, 7, \ldots$, construct the positive integers $1 + 2 \cdot 3$, $1 + 2 \cdot 3 \cdot 5$, $1 + 2 \cdot 3 \cdot 5 \cdot 7, \ldots$. Do you always get a prime number this way?

Middle-Level Problems

13. If p is an odd prime, show that p is of the form:
   (a) $4n + 1$ or $4n + 3$ for some n.
   (b) $6n + 1$ or $6n + 5$ for some n.

14. Adapt the proof of Theorem 1.5.9 to prove:
   (a) There is an infinite number of primes of the form $4n + 3$.
   (b) There is an infinite number of primes of the form $6n + 5$.

15. Show that no integer $u = 4n + 3$ can be written as $u = a^2 + b^2$, where a, b are integers.

16. If T is an infinite subset of $\mathbb{N}$, the set of all positive integers, show that there is a 1-1 mapping of T *onto* $\mathbb{N}$.

17. If p is a prime, prove that one cannot find nonzero integers a and b such that $a^2 = pb^2$. (This shows that $\sqrt{p}$ is *irrational*.)

6. MATHEMATICAL INDUCTION

If we look back at Section 5, we see that at several places—for instance, in the proof of Theorem 1.5.6—we say "argue as above and continue." This is not very satisfactory as a means of nailing down an argument.

What is clear is that we need some technique of avoiding such phrases when we want to prove a proposition about *all* the positive integers. This is provided for us by the *Principle of Mathematical Induction*; in fact, this will be the usual method that we shall use for proving theorems about all the positive integers.

> **Theorem 1.6.1.**   Let $P(n)$ be a statement about the positive integers such that:
>   (a) $P(1)$ is true.
>   (b) If $P(k)$ happens to be true for some integer $k \geq 1$,
>   then $P(k + 1)$ is also true.
> Then $P(n)$ is true for all $n \geq 1$.

Proof.   Actually, the arguments given in proving Theorems 1.5.7 and 1.5.8 are a prototype of the argument we give here.

Suppose that the theorem is false; then, by well-ordering, there is a least integer $m \geq 1$ for which $P(m)$ is not true. Since $P(1)$ is true, $m \neq 1$, hence $m > 1$. Now $1 \leq m - 1 < m$, so by the choice of m, $P(m - 1)$ must be valid. But then by the *inductive hypothesis* [Part (b)] we must have that $P(m)$ is true. This contradicts that $P(m)$ is not true. Thus there can be no integer for which P is not true, and so the theorem is proved. $\square$

We illustrate how to use induction with some rather diverse examples.

EXAMPLES

1. Suppose that n tennis balls are put in a straight line, touching each other. Then we claim that these balls make $n - 1$ contacts.

Proof.   If $n = 2$, the matter is clear. If for k balls we have $k - 1$ contacts, then adding one ball (on a line) adds one contact. So $k + 1$ balls would have k contacts. So if $P(n)$ is what is stated above about the tennis balls, we see that if $P(k)$ happens to be true, then so is $P(k + 1)$. Thus, by the theorem, $P(n)$ is true for all $n \geq 1$. $\square$

2. If p is a prime and $p | a_1 a_2 \cdots a_n$, then $p | a_i$ for some $1 \leq i \leq n$.

Proof.   Let $P(n)$ be the statement in Example 2. Then $P(1)$ is true, for if $p|a_1$, it certainly divides a_i for some $1 \leq i \leq 1$.

Suppose we know that $P(k)$ is true, and that $p|a_1a_2 \cdots a_ka_{k+1}$. Thus, by Theorem 1.5.6, since $p|(a_1a_2 \cdots a_k)a_{k+1}$ either $p|a_{k+1}$ (a desired conclusion) or $p|a_1 \cdots a_k$. In this second possibility, since $P(k)$ is true we have that $p|a_i$ for some $1 \leq i \leq k$. Combining both possibilities, we get that $p|a_j$ for some $1 \leq j \leq k + 1$. So Part (b) of Theorem 1.6.1 holds; hence $P(n)$ is true for all $n \geq 1$. $\square$

3. For $n \geq 1, 1 + 2 + \cdots + n = \frac{1}{2}n(n + 1)$.

Proof.   If $P(n)$ is the proposition that $1 + 2 + \cdots + n = \frac{1}{2}n(n + 1)$, then $P(1)$ is certainly true, for $1 = \frac{1}{2}(1 + 1)$. If $P(k)$ should be true, this means that

$$1 + 2 + \cdots + k = \frac{1}{2}k(k + 1).$$

The question is: Is $P(k + 1)$ then also true, that is, is $1 + 2 + \cdots + k + (k + 1) = \frac{1}{2}(k + 1)((k + 1) + 1)$? Now $1 + 2 + \cdots + k + (k + 1) = (1 + 2 + \cdots k) + (k + 1) = \frac{1}{2}k(k + 1) + (k + 1)$, since $P(k)$ is valid. But $\frac{1}{2}k(k + 1) + (k + 1) = \frac{1}{2}(k(k + 1) + 2(k + 1)) = \frac{1}{2}(k + 1)(k + 2)$, which assures us that $P(k + 1)$ is true. Thus the proposition $1 + 2 + \cdots + n = \frac{1}{2}n(n + 1)$ is true for all $n \geq 1$. $\square$

We must emphasize one point here: Mathematical induction is *not* a method for finding results about integers; it is a means of verifying a result. We could, by other means, find the formula given above for $1 + 2 + \cdots + n$.

Part (b) of Theorem 1.6.1 is usually called the *induction step*.

In the problems we shall give some other versions of the principle of induction.

Problems

Easier Problems

1. Prove that $1^2 + 2^2 + 3^2 + \cdots + n^2 = \frac{1}{6}n(n + 1)(2n + 1)$ by induction.

2. Prove that $1^3 + 2^3 + \cdots + n^3 = \frac{1}{4}n^2(n + 1)^2$ by induction.

3. Prove that a set having $n \geq 2$ elements has $\frac{1}{2}n(n-1)$ subsets having exactly two elements.

4. Prove that a set having $n \geq 3$ elements has $n(n-1)(n-2)/3!$ subsets having exactly three elements.

5. If $n \geq 4$ and S is a set having n elements, guess (from Problems 3 and 4) how many subsets having exactly 4 elements there are in S. Then verify your guess using mathematical induction.

*6. Complete the proof of Theorem 1.5.6, replacing the last sentence by an induction argument.

7. If $a > 1$, prove that $1 + a + a^2 + \cdots + a^n = (a^{n+1} - 1)/(a - 1)$ by induction.

8. By induction, show that

$$\frac{1}{1 \cdot 2} + \frac{1}{2 \cdot 3} + \cdots + \frac{1}{n(n+1)} = \frac{n}{n+1}.$$

9. Suppose that $P(n)$ is a proposition about the integers such that $P(n_0)$ is valid, and if $P(k)$ is true, so must $P(k + 1)$ be. What can you say about $P(n)$? Prove your statement.

10. Let $P(n)$ be a proposition about integers such that $P(1)$ is true and such that if $P(j)$ is true for all positive integers $j < k$, then $P(k)$ is true. Then prove that $P(n)$ is true for all positive integers n.

Middle-Level Problems

11. Give an example of a proposition that is *not* true for any positive integer, yet for which the induction step [Part (b) of Theorem 1.6.1] holds.

12. Prove by induction that a set having n elements has exactly 2^n subsets.

13. Prove by induction on n that $n^3 - n$ is always divisible by 3.

14. Using induction on n, generalize the result in Problem 13 to: If p is a prime number, then $n^p - n$ is always divisible by p. (**Hint:** The binomial theorem.)

15. Prove by induction that for a set having n elements the number of 1-1 mappings of this set onto itself is $n!$.

7. COMPLEX NUMBERS

We all know something about the integers, rational numbers, and real numbers—indeed, this assumption has been made for some of the text material and many of the problems have referred to these numbers. Unfortunately, the complex numbers and their properties are much less known to present-day college students. At one time the complex numbers were a part of the high school curriculum and the early college one. This is no longer the case. So we shall do a rapid development of this very important mathematical set.

The set of *complex numbers*, $\mathbb{C}$, is the set of all $a + bi$, where a, b are real and where we *declare*:

1. $a + bi = c + di$, a, b, c, d real, if and only if $a = c$ and $b = d$.
2. $(a + bi) \pm (c + di) = (a \pm c) + (b \pm d)i$.
3. $(a + bi)(c + di) = (ac - bd) + (ad + bc)i$.

This last property—multiplication—can best be remembered by using $i^2 = -1$ and multiplying out formally with this relation in mind.

For the complex number $z = a + bi$, a is called the *real part* of z and b the *imaginary part* of z. If a is 0, we call z *purely imaginary*.

We shall write $0 + 0i$ as 0 and $a + 0i$ as a. Note that $z + 0 = z$, $z1 = z$ for any complex number z.

Given $z = a + bi$, there is a complex number related to z, which we write as $\bar{z}$, defined by $\bar{z} = a - bi$. This complex number, $\bar{z}$, is called the *complex conjugate* of z. Taking this "—" gives us a mapping of $\mathbb{C}$ onto itself. We claim

Lemma 1.7.1. If $z, w \in \mathbb{C}$, then:
  (a) $\overline{(\bar{z})} = z$.
  (b) $\overline{(z + w)} = \bar{z} + \bar{w}$.
  (c) $\overline{(zw)} = \bar{z}\,\bar{w}$.
  (d) $z\bar{z}$ is real and nonnegative and is, in fact, positive if $z \neq 0$.
  (e) $z + \bar{z}$ is twice the real part of z.
  (f) $z - \bar{z}$ is twice the imaginary part of z times i.

Proof. Most of the parts of this lemma are straightforward and merely involve using the definition of complex conjugate. We do verify Parts (c) and (d).

Suppose that $z = a + bi$, $w = c + di$, where a, b, c, d are real. So $zw = (ac - bd) + (ad + bc)i$, hence

$$\overline{(zw)} = \overline{(ac - bd) + (ad + bc)i} = (ac - bd) - (ad + bc)i.$$

On the other hand, $\bar{z} = a - bi$ and $\bar{w} = c - di$, hence, by the definition of the product in $\mathbb{C}$, $\bar{z}\bar{w} = (ac - bd) - (ad + bc)i$. Comparing this with the result that we obtained for $\overline{(zw)}$, we see that indeed $\overline{(zw)} = \bar{z}\bar{w}$. This verifies Part (c).

We go next to the proof of Part (d). Suppose that $z = a + bi \neq 0$; then $\bar{z} = a - bi$ and $z\bar{z} = a^2 + b^2$. Since a, b are real and not both 0, $a^2 + b^2$ is real and positive, as asserted in Part (d). $\square$

The proof of Part (d) of Lemma 1.7.1 shows that if $z = a + bi \neq 0$, then $z\bar{z} = a^2 + b^2 \neq 0$ and $z(\bar{z}/(a^2 + b^2)) = 1$, so

$$\frac{\bar{z}}{a^2 + b^2} = \frac{a}{a^2 + b^2} - \left(\frac{b}{a^2 + b^2}\right)i$$

acts like the inverse $1/z$ of z. This allows us to carry out division in $\mathbb{C}$, staying in $\mathbb{C}$ while doing so.

We now list a few properties of $\mathbb{C}$.

Lemma 1.7.2. $\mathbb{C}$ behaves under its sum and product according to the following: If $u, v, w \in \mathbb{C}$, then
(a) $u + v = v + u$.
(b) $(u + v) + w = u + (v + w)$.
(c) $uv = vu$.
(d) $(uv)w = u(vw)$.
(e) $u \neq 0$ implies that $u^{-1} = 1/u$ exists in $\mathbb{C}$ such that $uu^{-1} = 1$.

Proof. We leave the proofs of these various parts to the reader. $\square$

These properties of $\mathbb{C}$ make of $\mathbb{C}$ what we shall call a *field*, which we shall study in much greater depth later in the book. What the lemma says is that we are allowed to calculate in $\mathbf{C}$ more or less as we did with real numbers. However, $\mathbf{C}$ is a much richer structure than is the set of real numbers.

We now introduce a "size" function on $\mathbf{C}$.

> **Definition.** If $z = a + bi \in \mathbf{C}$, then the *absolute value* of z, written as $|z|$, is defined by $|z| = \sqrt{z\bar{z}} = \sqrt{a^2 + b^2}$.

We shall see, in a few moments, what this last definition means geometrically. In the meantime we prove

> **Lemma 1.7.3.** If $u, v \in \mathbf{C}$, then $|uv| = |u|\,|v|$.

Proof. By definition, $|u| = \sqrt{u\bar{u}}$ and $|v| = \sqrt{v\bar{v}}$. Now

$$|uv| = \sqrt{(uv)\overline{(uv)}} = \sqrt{(uv)(\bar{u}\bar{v})} \qquad \text{[by Part (c) of Lemma 1.7.1]}$$

$$= \sqrt{(u\bar{u})(v\bar{v})} \qquad \text{(by Lemma 1.7.2)}$$

$$= \sqrt{u\bar{u}}\,\sqrt{v\bar{v}} = |u|\,|v|. \quad \square$$

Another way of verifying this lemma is to write $u = a + bi$, $v = c + di$, $uv = (ac - bd) + (ad + bc)i$ and to note the identity

$$(ac - bd)^2 + (ad + bc)^2 = (a^2 + b^2)(c^2 + d^2).$$

Note several small points about conjugates. If $z \in \mathbf{C}$, then z is real if and only if $\bar{z} = z$, and z is purely imaginary if and only if $\bar{z} = -z$. If $z, w \in \mathbf{C}$, then

$$\overline{(z\bar{w} + \bar{z}w)} = \bar{z}\bar{\bar{w}} + \bar{\bar{z}}\bar{w} = \bar{z}w + z\bar{w},$$

so $z\bar{w} + \bar{z}w$ is real. We want to get an upper bound for $|z\bar{w} + \bar{z}w|$; this will come up in the proof of Theorem 1.7.5 below.

But first we must digress for a moment to obtain a statement about quadratic expressions.

Lemma 1.7.4. If a, b, c are real, and $a > 0$ are such that $a\alpha^2 + b\alpha + c \geq 0$ for every real α, then $b^2 - 4ac \leq 0$.

Proof. Consider the quadratic expression for $\alpha = -b/2a$. We get $a(-b/2a)^2 + b(-b/2a) + c \geq 0$. Simplifying this, we obtain that $(4ac - b^2)/4a \geq 0$, and since $a > 0$, we end up with $4ac - b^2 \geq 0$, and so $b^2 - 4ac \leq 0$. $\square$

We use this result immediately to prove the important

Theorem 1.7.5 (Triangle Inequality). For $z, w \in \mathbb{C}$, $|z + w| \leq |z| + |w|$.

Proof. If $z = 0$, there is nothing to prove, so we may assume that $z \neq 0$; thus $z\bar{z} > 0$. Now, for α real,

$$0 \leq |\alpha z + w|^2 = (\alpha z + w)\overline{(\alpha z + w)} = (\alpha z + w)(\alpha \bar{z} + \bar{w})$$

$$= \alpha^2 z\bar{z} + \alpha(z\bar{w} + \bar{z}w) + w\bar{w}.$$

If $a = z\bar{z} > 0$, $b = z\bar{w} + \bar{z}w$, $c = w\bar{w}$, then Lemma 1.7.4 tells us that $b^2 - 4ac = (z\bar{w} + \bar{z}w)^2 - 4(z\bar{z})(w\bar{w}) \leq 0$, hence $(z\bar{w} + \bar{z}w)^2 \leq 4(z\bar{z})(w\bar{w}) = 4|z|^2|w|^2$. Therefore, $z\bar{w} + \bar{z}w \leq 2|z||w|$.

For $\alpha = 1$ above,

$$|z + w|^2 = z\bar{z} + w\bar{w} + z\bar{w} + \bar{z}w = |z|^2 + |w|^2 + z\bar{w} + \bar{z}w$$

$$\leq |z|^2 + |w|^2 + 2|z||w|$$

from the result above. In other words, $|z + w|^2 \leq (|z| + |w|)^2$; taking square roots we get the desired result, $|z + w| \leq |z| + |w|$. $\square$

Why is this result called the triangle inequality? The reason will be clear once we view the complex numbers geometrically. Represent the complex number $z = a + bi$ as the point having coordinates (a, b) in the x-y plane.

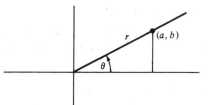

The distance r of this point from the origin is $\sqrt{a^2 + b^2}$, in other words, $|z|$. The angle θ is called the *argument* of z and, as we see, $\tan \theta = b/a$. Also, $a = r \cos \theta$, $b = r \sin \theta$; therefore, $z = a + bi = r(\cos \theta + i \sin \theta)$. This representation of z is called its *polar form*.

Given $z = a + bi$, $w = c + di$, then their sum is $z + w = (a + c) + (b + d)i$; geometrically, we have the picture

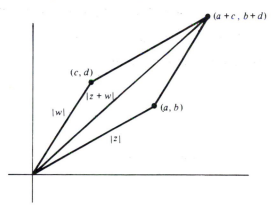

The statement $|z + w| \leq |z| + |w|$ merely reflects that in the triangle T one side is of smaller length than the sum of the lengths of the other two sides: thus the term *triangle inequality*.

The complex numbers of the form $\cos \theta + i \sin \theta$ that come up in the polar form are very interesting numbers indeed. First, notice that

$$|\cos \theta + i \sin \theta| = \sqrt{\cos^2 \theta + \sin^2 \theta} = \sqrt{1} = 1,$$

so they give us many complex numbers of absolute value 1. In truth they give us *all* the complex numbers of absolute value 1; to see this just go back and look at the polar form of such a number.

Let's recall two basic identities from trigonometry, $\cos(\theta + \psi) = \cos\theta\cos\psi - \sin\theta\sin\psi$ and $\sin(\theta + \psi) = \sin\theta\cos\psi + \cos\theta\sin\psi$. Therefore,

$$(\cos\theta + i\sin\theta)(\cos\psi + i\sin\psi)$$

$$= (\cos\theta\cos\psi - \sin\theta\sin\psi) + i(\sin\theta\cos\psi + \cos\theta\sin\psi)$$

$$= \cos(\theta + \psi) + i\sin(\theta + \psi).$$

Thus, in multiplying two complex numbers, the argument of the product is the sum of the arguments of the factors.

This has another very interesting consequence.

Theorem 1.7.6 (De Moivre's Theorem). For any integer $n \geq 1$, $(\cos\theta + i\sin\theta)^n = \cos(n\theta) + i\sin(n\theta)$.

Proof. We proceed by induction on n. If $n = 1$, the statement is obviously true. Assume then that for some k, $(\cos\theta + i\sin\theta)^k = \cos k\theta + i\sin k\theta$. Thus

$$(\cos\theta + i\sin\theta)^{k+1} = (\cos\theta + i\sin\theta)^k(\cos\theta + i\sin\theta)$$

$$= (\cos k\theta + i\sin k\theta)(\cos\theta + i\sin\theta)$$

$$= \cos(k+1)\theta + i\sin(k+1)\theta,$$

by the result of the paragraph above. This completes the induction step; hence the result is true for all integers $n \geq 1$. $\square$

In the problems we shall see that De Moivre's Theorem is true for all integers m; in fact, it is true even if m is rational.

Consider the following special case:

$$\theta_n = \cos\frac{2\pi}{n} + i\sin\frac{2\pi}{n}, \qquad \text{where } n \geq 1 \text{ is an integer.}$$

By De Moivre's Theorem,

$$\left(\cos\left(\frac{2\pi}{n}\right) + i\sin\left(\frac{2\pi}{n}\right)\right)^n$$

$$= \cos\left(n\left(\frac{2\pi}{n}\right)\right) + i\sin\left(n\left(\frac{2\pi}{n}\right)\right)$$

$$= \cos 2\pi + i\sin 2\pi = 1.$$

So $\theta_n^n = 1$; you can verify that $\theta_n^m \neq 1$ if $0 < m < n$. This number θ_n is called a *primitive n*th *root of unity*.

Problems

1. Multiply.
   (a) $(6 - 7i)(8 + i)$.
   (b) $(\frac{2}{3} + \frac{3}{2}i)(\frac{2}{3} - \frac{3}{2}i)$.
   (c) $(6 + 7i)(8 - i)$.

2. Express z^{-1} in the form $z^{-1} = a + bi$ for:
   (a) $z = 6 + 8i$.
   (b) $z = 6 - 8i$.
   (c) $z = \dfrac{1}{\sqrt{2}} + \dfrac{1}{\sqrt{2}}i$.

*3. Show that $(\bar{z})^{-1} = \overline{(z^{-1})}$.

4. Find $(\cos\theta + i\sin\theta)^{-1}$.

5. Verify all parts of Lemma 1.7.1.

*6. Show that z is real if and only if $\bar{z} = z$, and is purely imaginary if and only if $\bar{z} = -z$.

7. Verify the commutative law of multiplication $zw = wz$ in $\mathbb{C}$.

8. Show that for $z \neq 0$, $|z^{-1}| = 1/|z|$.

9. Find:
   (a) $|6 - 4i|$.
   (b) $|\frac{1}{2} + \frac{2}{3}i|$.
   (c) $\left| \dfrac{1}{\sqrt{2}} + \dfrac{1}{\sqrt{2}}i \right|$.

10. Show that $|\bar{z}| = |z|$.

11. Find the polar form for
    (a) $z = \dfrac{\sqrt{2}}{2} - \dfrac{1}{\sqrt{2}}i$.
    (b) $z = 4i$.
    (c) $z = \dfrac{6}{\sqrt{2}} + \dfrac{6}{\sqrt{2}}i$.
    (d) $z = -\dfrac{13}{2} + \dfrac{39}{2\sqrt{3}}i$.

12. Prove that $(\cos(\frac{1}{2}\theta) + i\sin(\frac{1}{2}\theta))^2 = \cos\theta + i\sin\theta$.

13. By direct multiplication show that $(\frac{1}{2} + \frac{1}{2}\sqrt{3}\,i)^3 = -1$.

Middle-Level Problems

14. Show that $(\cos\theta + i\sin\theta)^m = \cos(m\theta) + i\sin(m\theta)$ for *all* integers *m*.

15. Show that $(\cos\theta + i\sin\theta)^r = \cos(r\theta) + i\sin(r\theta)$ for all rational numbers *r*.

16. If $z \in \mathbf{C}$ and $n \geq 1$ is any positive integer, show that there are *n* distinct complex numbers *w* such that $z = w^n$.

17. Find the necessary and sufficient condition on *k* such that:

    (a) $\left(\cos\left(\dfrac{2\pi k}{n} \right) + i\sin\left(\dfrac{2\pi k}{n} \right) \right)^n = 1$.

    (b) $\left(\cos\left(\dfrac{2\pi k}{n} \right) + i\sin\left(\dfrac{2\pi k}{n} \right) \right)^m \neq 1$ if $0 < m < n$.

18. Viewing the x-y plane as the set of all complex numbers $x + iy$, show that multiplication by i induces a 90° rotation of the x-y plane in a counterclockwise direction.

19. In Problem 18, interpret geometrically what multiplication by the complex number $a + bi$ does to the x-y plane.

20. Prove that $|z + w|^2 + |z - w|^2 = 2(|z|^2 + |w|^2)$.

*21. Consider the set $A = \{a + bi | a, b \in Z\}$. Prove that there is a 1-1 correspondence of A onto N. (A is called the set of *Gaussian integers*.)

22. If a is a (complex) root of the polynomial

$$x^n + \alpha_1 x^{n-1} + \cdots + \alpha_{n-1} x + \alpha_n,$$

where the α_i are real, show that $\bar{a}$ must also be a root. [r is a root of a polynomial $p(x)$ if $p(r) = 0$.]

Harder Problems

23. Find the necessary and sufficient conditions on z and w in order that $|z + w| = |z| + |w|$.

24. Do Problem 23 for $|z_1 + \cdots + z_k| = |z_1| + \cdots + |z_k|$.

*25. The complex number θ is said to have *order* $n \geq 1$ if $\theta^n = 1$, and $\theta^m \neq 1$ for $0 < m < n$. Show that if θ has order n and $\theta^k = 1$, where $k > 0$, then $n | k$.

*26. Find all complex numbers θ having order n. (These are the *primitive nth roots of unity*.)

CHAPTER 2

Groups

1. DEFINITIONS AND EXAMPLES OF GROUPS

We have seen in Section 4 of Chapter 1 that given any nonempty set, the set $A(S)$ of all 1-1 mappings of S onto itself is not just a set alone, but has a far richer texture. The possibility of combining two elements of $A(S)$ to get yet another element of $A(S)$ endows $A(S)$ with an algebraic structure. We recall how this was done: If $f, g \in A(S)$, then we combine them to form the mapping fg defined by $(fg)(s) = f(g(s))$ for every $s \in S$. We called fg the *product* of f and g, and verified that $fg \in A(S)$, and that this product obeyed certain rules. From the myriad of possibilities we somehow selected four particular rules that govern the behavior of $A(S)$ relative to this product.

These four rules were

1. *Closure*, namely if $f, g \in A(S)$, then $fg \in A(S)$. We say that $A(S)$ is *closed* under this product.
2. *Associativity*, that is, given $f, g, h \in A(S)$, then $f(gh) = (fg)h$.
3. *Existence of a unit element*, namely, there exists a particular element $i \in A(S)$ (the identity mapping) such that $fi = if = f$ for all $f \in A(S)$.

47

4. *Existence of inverses*, that is, given $f \in A(S)$ there exists an element, denoted by f^{-1}, in $A(S)$ such that $ff^{-1} = f^{-1}f = i$.

To justify or motivate why these four specific attributes of $A(S)$ were singled out, in contradistinction to some other set of properties, is not easy to do. In fact, in the history of the subject it took quite some time to recognize that these four properties played the key role. We have the advantage of historical hindsight, and with this hindsight we choose them not only to study $A(S)$, but also as the chief guidelines for abstracting to a much wider context.

Although we saw that the four properties above enabled us to calculate concretely in $A(S)$, there were some differences with the kind of calculations we are used to. If S has three or more elements, we saw in Problem 15, Chapter 1, Section 3 that it is possible for $f, g \in A(S)$ to have $fg \neq gf$. However, this did not present us with insurmountable difficulties.

Without any further polemic we go to the

Definition.   A nonempty set G is said to be a *group* if in G there is defined an operation $*$ such that:

(a) $a, b \in G$ implies that $a * b \in G$.
   (We describe this by saying that G is *closed* under $*$.)
(b) Given $a, b, c \in G$, then $a * (b * c) = (a * b) * c$.
   (This is described by saying that the *associative law* holds in G.)
(c) There exists a special element $e \in G$ such that $a * e = e * a = a$ for all $a \in G$ (e is called the *identity* or *unit element* of G).
(d) For every $a \in G$ there exists an element $b \in G$ such that $a * b = b * a = e$. (We write this element b as a^{-1} and call it the *inverse* of a in G.)

These four defining postulates (called the *group axioms*) for a group were, after all, patterned after those that hold in $A(S)$. So it is not surprising that $A(S)$ is a group relative to the operation "composition of mappings."

The operation ∗ in G is usually called the *product*, but keep in mind that this has nothing to do with product as we know it for the integers, rationals, reals, or complexes. In fact, as we shall see below, in many familiar examples of groups that come from numbers, what we call the product in these groups is actually the addition of numbers. *However, a general group need have no relation whatsoever to a set of numbers.* We reiterate: A group is no more, no less, than a set satisfying the four group axioms.

Before starting to look into the nature of groups, we look at some examples.

EXAMPLES OF GROUPS

1. Let $\mathbb{Z}$ be the set of all integers and let ∗ be the ordinary addition, $+$, in $\mathbb{Z}$. That $\mathbb{Z}$ is closed and associative under ∗ are basic properties of the integers. What serves as the unit element, e, of $\mathbb{Z}$ under ∗? Clearly, since $a = a \ast e = a + e$, we have $e = 0$, and 0 is the required identity element under addition. What about a^{-1}? Here too, since $e = 0 = a \ast a^{-1} = a + a^{-1}$, the a^{-1} in this instance is $-a$, and clearly $a \ast (-a) = a + (-a) = 0$.

2. Let $\mathbb{Q}$ be the set of all rational numbers and let the operation ∗ on $\mathbb{Q}$ be the ordinary addition of rational numbers. As above, $\mathbb{Q}$ is easily shown to be a group under ∗. Note that $\mathbb{Z} \subset \mathbb{Q}$ and both $\mathbb{Z}$ and $\mathbb{Q}$ are groups under the same operation ∗.

3. Let $\mathbb{Q}'$ be the set of all *nonzero* rational numbers and let the operation ∗ on $\mathbb{Q}'$ be the ordinary multiplication of rational numbers. By the familiar properties of the rational numbers we see that $\mathbb{Q}'$ forms a group relative to ∗.

4. Let $\mathbb{R}'$ be the set of all *positive real* numbers and let the operation ∗ on $\mathbb{R}'$ be the ordinary product of real numbers. Again it is easy to check that $\mathbb{R}'$ is a group under ∗.

5. Let E_n be the powers θ_n^i, $i = 0, 1, 2, \ldots, n - 1$, where θ_n is the complex number $\theta_n = \cos(2\pi/n) + i \sin(2\pi/n)$. Let $\theta_n^k \ast \theta_n^j = \theta_n^{k+j}$, the ordinary product of the powers of θ as complex numbers. By De Moivre's Theorem we saw that $\theta_n^n = 1$. We leave it to the reader to verify that E_n is a group under ∗.

Note one striking difference between the Examples 1 to 4 and Example 5; the first four have an infinite number of elements, whereas U_n has a finite number, n, of elements.

> **Definition.** A group G is said to be a *finite group* if it has a finite number of elements. The number of elements in G is called the *order* of G and is denoted by $|G|$.

Thus E_n above is a finite group, and $|E_n| = n$.

All the examples presented above satisfy the additional property that $a * b = b * a$ for any pairs of elements. This need not be true in a group; just witness the case of $A(S)$, where S has three or more elements; there we saw that we could find $f, g \in A(S)$ such that $fg \neq gf$.

This prompts us to single out as special those groups of G in which $a * b = b * a$ for all $a, b \in G$.

> **Definition.** A group G is said to be abelian *if $a * b = b * a$ for all $a, b \in G$.*

The word abelian derives from the name of the great Norwegian mathematician Niels Henrik Abel (1802–1829), one of the greatest scientists Norway has ever produced.

A group that is not abelian is called *nonabelian*, a not too surprising choice of name.

We give some examples of some nonabelian groups. Of course, the $A(S)$ afford us an infinite family of such. But we present a few other examples in which we can compute quite readily.

6. Let $\mathbb{R}$ be the set of all real numbers, and let G be the set of all mappings $T_{a,b} : \mathbb{R} \to \mathbb{R}$ defined by $T_{a,b}(r) = ar + b$ for any real number r, where a, b are real numbers and $a \neq 0$. Thus, for instance, $T_{5,-6}$ is such that $T_{5,-6}(r) = 5r - 6$; $T_{5,-6}(14) = 5 \cdot 14 - 6 = 64$, $T_{5,-6}(\pi) = 5\pi - 6$. The $T_{Ta,b}$ are 1-1 mappings of $\mathbb{R}$ onto itself, and we let $T_{a,b} * T_{c,d}$ be the product of these mappings. So

$$(T_{a,b} * T_{c,d})(r) = T_{a,b}(T_{c,d}(r)) = aT_{c,d}(r) + b = a(cr + d) + b$$

$$= (ac)r + (ad + b) = T_{ac,\,ad+b}(r).$$

So we have the formula

(1) $$T_{a,b} * T_{c,d} = T_{ac,ad+b}.$$

This result shows us that $T_{a,b} * T_{c,d}$ is in G—for it satisfies the membership requirement for belonging to G—so G is closed under $*$. Since we are talking about the product of mappings (i.e., the composition of mappings), $*$ is associative. The element $T_{1,0} = i$ is the identity mapping of $\mathbb{R}$ onto itself. Finally, what is $T_{a,b}^{-1}$? Can we find $x \neq 0$, y, real numbers, such that

$$T_{a,b} * T_{x,y} = T_{x,y} * T_{a,b} = T_{1,0}?$$

Go back to (1) above; we thus want $T_{ax,ay+b} = T_{1,0}$, that is, $ax = 1$, $ay + b = 0$. Remember now that $a \neq 0$, so if we put $x = a^{-1}$, $y = -a^{-1}b$, the required relations are satisfied. One verifies immediately that

$$T_{a,b} * T_{a^{-1}, -a^{-1}b} = T_{a^{-1}, -a^{-1}b} * T_{a,b} = T_{1,0}.$$

So G is indeed a group.

What is $T_{c,d} * T_{a,b}$? According to the formula given in (1), where we replace a by c, c by a, b by d, d by b, we get

(2) $$T_{c,d} * T_{a,b} = T_{ca,cb+d}.$$

Thus $T_{c,d} * T_{a,b} = T_{a,b} * T_{c,d}$ if and only if $bc + d = ad + b$. This fails to be true, for instance, if $a = 1$, $b = 1$, $c = 2$, $d = 3$. So G is nonabelian.

7. Let $H \subset G$, where G is the group in Example 6, defined by $H = \{T_{a,b} \in G | a \text{ is rational}, b \text{ any real}\}$. We leave it to the reader to verify that H is a group under the operation $*$ defined on G. H is nonabelian.

8. Let $K \subset H \subset G$, where H, G are as above and such that $K = \{T_{1,b} \in G | b \text{ any real}\}$. The reader should check that K is a group relative to the operation $*$ of G, and *that K is, however, abelian.*

9. Let S be the plane, that is, $S = \{(x, y) | x, y \text{ real}\}$ and consider $f, g \in A(S)$ defined by $f(x, y) = (-x, y)$ and $g(x, y) = (-y, x)$; f is the reflection about the y-axis and g is the rotation through $90°$ in a

counterclockwise direction about the origin. We then define $G = \{ f^i g^j | i = 0, 1; \; j = 0, 1, 2, 3 \}$, and let $*$ in G be the product of elements in $A(S)$. Clearly, $f^2 = g^4 =$ identity mapping;

$$(f * g)(x, y) = (fg)(x, y) = f(g(x, y)) = f(-y, x) = (y, x)$$

and

$$(g * f)(x, y) = g(f(x, y)) = g(-x, y) = (-y, -x).$$

So $g * f \neq f * g$. We leave it to the reader to verify that $g * f = f * g^{-1}$ and G is a nonabelian group of order 8. This group is called the *dihedral group* of order 8. [Try to find a formula for $(f^i g^j) * (f^s g^t) = f^a g^b$ that expresses a, b in terms of i, j, s, and t.]

10. Let S be as in Example 9 and f the mapping in Example 9. Let $n > 2$ and let h be the rotation of the plane about the origin through an angle of $2\pi/n$ in the counterclockwise direction. We then define $G\{ f^k h^j | k = 0, 1; \; j = 0, 1, 2, \ldots, n - 1 \}$ and define the product $*$ in G via the usual product of mappings. One can verify that $f^2 = h^n =$ identity mapping, and $fh = h^{-1}f$. These relations allow us to show (with some effort) that G is a nonabelian group of order $2n$. G is called the *dihedral group* of order $2n$.

11. Let $G = \{(f \in A(S) | f(s) \neq s$ for only a finite number of $s \in S\}$, where we suppose that S is an *infinite* set. We claim that G is a group under the product $*$ in $A(S)$. The associativity holds automatically in G, since it already holds in $A(S)$. Also, $i \in G$, since $i(s) = s$ for *all* $s \in S$. So we must show that G is closed under the product and if $f \in G$, then $f^{-1} \in G$.

We first dispose of the closure. Suppose that $f, g \in G$; then $f(s) = s$ except, say, for $s_1, s_2, \ldots, s_n$ and $g(s) = s$ except for $s_1', s_2', \ldots, s_m'$. Then $(fg)(s) = f(g(s)) = s$ for all s other than $s_1, s_2, \ldots, s_n, s_1', \ldots, s_m'$ (and possibly even for some of these). So fg moves only a finite number of elements of S, so $fg \in G$.

Finally, if $f(s) = s$ for all s other than $s_1, s_2, \ldots, s_n$, then $f^{-1}(f(s)) = f^{-1}(s)$, but $f^{-1}(s) = f^{-1}(f(s)) = (f^{-1}f)(s) = i(s) = s$. So we obtain that $f^{-1}(s) = s$ for all s except $s_1, \ldots, s_n$. Thus $f^{-1} \in G$ and G satisfies all the group axioms, hence G is a group.

12. Let G be the set of all mappings T_θ, where T_θ is the rotation of a given circle about its center through an angle θ in the clockwise

direction. In G define $*$ by the composition of mappings. Since, as is readily verified, $T_\theta * T_\psi = T_{\theta+\psi}$, G is closed under $*$. The other group axioms check out easily. Note that $T_{2\pi} = T_0 =$ the identity mapping, and $T_\theta^{-1} = T_{-\theta} = T_{2\pi-\theta}$. G is an abelian group.

As we did for $A(S)$ we introduce the shorthand notation a^n for

$$\underbrace{a * a * a \cdots * a}_{n \text{ times}}$$

and define $a^{-n} = (a^{-1})^n$, for n a positive integer, and $a^0 = e$. The usual rules of exponents then hold, that is, $(a^m)^n = a^{mn}$ and $a^m * a^n = a^{m+n}$ for *any* integers m and n.

Note that with this notation, if G is the group of integers under $+$, then a^n is really na.

Having seen the 12 examples of groups above, the reader might get the impression that all, or almost all, sets with some operation $*$ form groups. This is far from true. We now give some examples of nongroups. In each case we check the four group axioms and see which of these fail to hold.

NONEXAMPLES

1. Let G be the set of all integers, and let $*$ be the ordinary product of integers in G. Since $a * b = ab$, for $a, b \in G$ we clearly have that G is closed and associative relative to $*$. Furthermore, the number 1 serves as the unit element, since $a * 1 = a1 = a = 1a = 1 * a$ for every $a \in G$. So we are three-fourths of the way to proving that G is a group. All we need is inverses for the elements of G, relative to $*$, to lie in G. But this just isn't so. Clearly, we cannot find an integer b such that $0 * b = 0b = 1$, since $0b = 0$ for all b. But even other integers fail to have inverses in G. For instance, we *cannot find an integer* b such that $3 * b = 1$ (for this would require that $b = \frac{1}{3}$, and $\frac{1}{3}$ is not an integer).

2. Let G be the set of all nonzero real numbers and define, for $a, b \in G$, $a * b = a^2b$; thus $4 * 5 = 4^2(5) = 80$. Which of the group axioms hold in G under this operation $*$ and which fail to hold? Certainly, G is closed under $*$. Is $*$ associative? If so, $(a * b) * c = a * (b * c)$, that is, is $(a * b)^2c = a^2(b * c)$, and so $(a^2b)^2c = a^2(b^2c)$,

which boils down to $a^2 = 1$, which holds only for $a = \pm 1$. So, in general, the associative law does *not* hold in G relative to $*$. We similarly can verify that G does not have a unit element. Thus even to discuss inverses relative to $*$ would not make sense.

3. Let G be the set of all *positive* integers, under $*$ where $a * b = ab$, the ordinary product of integers. Then one can easily verify that G fails to be a group *only because it fails to* have inverses for some (in fact, most) of its elements relative to $*$.

We shall find some other nonexamples of groups in the exercises.

Problems

Easier Problems

1. Determine if the following sets G with the operation indicated form a group. If not, point out which of the group axioms fail.
   (a) G = set of all integers, $a * b = a - b$.
   (b) G = set of all integers, $a * b = a + b + ab$.
   (c) G = set of nonnegative integers, $a * b = a + b$.
   (d) G = set of all rational numbers $\neq -1$, $a * b = a + b + ab$.
   (e) G = set of all rational numbers with denominator divisible by 5 (written so that numerator and denominator are relatively prime), $a * b = a + b$.
   (f) G a set having more than one element, $a * b = a$ for all $a, b \in G$.

2. In the group G defined in Example 6, show that the set $H = \{T_{a,b} | a = \pm 1, b \text{ any real}\}$ forms a group under the $*$ of G.

3. Verify that Example 7 is indeed an example of a group.

4. Prove that K defined in Example 8 is an abelian group.

5. In Example 9, prove that $g * f = f * g^{-1}$, and that G is a group, is nonabelian, and is of order 8.

6. Let H and G be as in Examples 6 and 7, respectively. Show that if $T_{a,b} \in G$, then $T_{a,b} * V * T_{a,b}^{-1} \in H$ if $V \in H$.

7. Do the same as in Problem 6 for the group $K \subset G$ of Example 8.

8. If G is an abelian group, prove that $(a * b)^n = a^n * b^n$ for all integers n.

9. If G is a group in which $a^2 = e$ for all $a \in G$, show that G is abelian.

10. If G is the group in Example 6, find all $T_{a,b} \in G$ such that $T_{a,b} * T_{1,x} = T_{1,x} * T_{a,b}$ for *all* real x.

11. In Example 10, for $n = 3$ find a formula that expresses $(f^i h^j) * (f^s h^t)$ as $f^a h^b$. From this formula show that G is a nonabelian group of order 6.

12. Do Problem 11 for $n = 4$.

13. Show that any group of order 4 or less is abelian.

14. If G is any group and $a, b, c \in G$, show that if $a * b = a * c$, then $b = c$, and if $b * a = c * a$, then $b = c$.

15. Express $(a * b)^{-1}$ in terms of a^{-1} and b^{-1}.

16. Using the result of Problem 15, prove that a group G in which $a = a^{-1}$ for every $a \in G$ must be abelian.

17. In any group G, prove that $(a^{-1})^{-1} = a$ for all $a \in G$.

18. If G is a finite group of *even* order, show that there must be an element $a \neq e$ such that $a = a^{-1}$. (**Hint:** Try to use the result of Problem 17.)

19. In S_3, show that there are four elements x satisfying $x^2 = e$ and three elements y satisfying $y^3 = e$.

20. Find all the elements in S_4 such that $x^4 = e$.

Middle-Level Problems

21. Show that a group of order 5 must be abelian.

22. Show that the set defined in Example 10 is a group, is nonabelian, and has order $2n$. Do this by finding the formula for $(f^i h^j) * (f^s h^t)$ in the form $f^a h^b$.

23. In the group G of Example 6, find all elements $U \in G$ such that $U * T_{a,b} = T_{a,b} * U$ for every $T_{a,b} \in G$.

24. If G is the dihedral group of order $2n$ as defined in Example 10, prove that:
    (a) If n is odd and $a \in G$ is such that $a * b = b * a$ for all $b \in G$, then $a = e$.
    (b) If n is even, show that there is an $a \in G$, $a \neq e$, such that $a * b = b * a$ for all $b \in G$.
    (c) If n is even, find all the elements $a \in G$ such that $a * b = b * a$ for all $b \in G$.

25. If G is any group, show that:
    (a) e is unique (i.e., if $f \in G$ also acts as a unit element for G, then $f = e$).
    (b) Given $a \in G$, then $a^{-1} \in G$ is unique.

26. If G is a finite group, prove that, given $a \in G$, there is a positive integer n, depending on a, such that $a^n = e$.

27. In Problem 26, show that there is an integer $m > 0$ such that $a^m = e$ for *all* $a \in G$.

Harder Problems

28. Let G be a set with an operation $*$ such that:
    1. G is closed under $*$.
    2. $*$ is associative.
    3. There exists an element $e \in G$ such that $e * x = x$ for all $x \in G$.
    4. Given $x \in G$, there exists a $y \in G$ such that $y * x = e$.
    Prove that G is a group. (Thus you must show that $x * e = x$ and $x * y = e$ for e, y as above.)

29. Let G be a *finite* set with an operation $*$ such that:
    1. G is closed under $*$.
    2. $*$ is associative.
    3. Given $a, b, c \in G$ with $a * b = a * c$, then $b = c$.
    4. Given $a, b, c \in G$ with $b * a = c * a$, then $b = c$.
    Prove that G must be a group under $*$.

30. Give an example to show that the result of Problem 29 can be false if G is an infinite set.

31. Let G be the group of all nonzero real numbers under the operation $*$ which is the ordinary multiplication of real numbers, and let H be the group of all real numbers under the operation $\#$, which is the addition of real numbers.

(a) Show that there is a mapping $F: G \to H$ of G onto H which satisfies $F(a*b) = F(a)\#F(b)$ for all $a, b \in G$ [i.e., $F(ab) = F(a) + F(b)$].

(b) Show that no such mapping F can be 1-1.

2. SOME SIMPLE REMARKS

In this short section we show that certain formal properties which follow from the group axioms hold in any group. As a matter of fact, most of these results have already occurred as problems at the end of the preceding section.

It is a little clumsy to keep writing the $*$ for the product in G, and *from now on we shall write the product $a * b$ simply as ab for all $a, b \in G$.*

The first such formal results we prove are contained in

Lemma 2.2.1.   If G is a group, then:

(a) Its identity element is *unique*.

(b) Every $a \in G$ has a *unique* inverse $a^{-1} \in G$.

(c) If $a \in G$, $(a^{-1})^{-1} = a$.

(d) For $a, b \in G$, $(ab)^{-1} = b^{-1}a^{-1}$.

Proof.   We start with Part (a). What is expected of us to carry out the proof? We must show that if $e, f \in G$ and $af = fa = a$ for all $a \in G$ and $ae = ea = a$ for all $a \in G$, then $e = f$. This is very easy, for then $e = ef$ and $f = ef$; hence $e = ef = f$, as required.

Instead of proving Part (b), we shall prove a stronger result (listed below as Lemma 2.2.2), which will have Part (b) as an immediate consequence. We claim that in a group G if $ab = ac$, then $b = c$; that is, we can *cancel a given element from the same side of an equation*. To see this, we have, for $a \in G$, an element $u \in G$ such that $ua = e$. Thus from $ab = ac$ we have

$$u(ab) = u(ac),$$

so, by the associative law, $(ua)b = (ua)c$, that is, $eb = ec$. Hence $b = eb = ec = c$, and our result is established. A similar argument shows that if $ba = ca$, then $b = c$. However, we *cannot conclude* from $ab = ca$ that $b = c$; in any abelian group, yes, but in general, no.

Now to get Part (b) as an implication of the cancellation result. Suppose that $b, c \in G$ act as inverses for a; then $ab = e = ac$, so by cancellation $b = c$ and we see that the inverse of a is unique. We shall always write it as a^{-1}.

To see Part (c), note that by definition $a^{-1}(a^{-1})^{-1} = e$; but $a^{-1}a = e$, so by cancellation in $a^{-1}(a^{-1})^{-1} = e = a^{-1}a$ we get that $(a^{-1})^{-1} = a$.

Finally, for Part (d) we calculate

$$(ab)(b^{-1}a^{-1}) = ((ab)b^{-1})a^{-1} \quad \text{(associative law)}$$

$$= (a(bb^{-1})a^{-1} \quad \text{(again the associative law)}$$

$$= (ae)a^{-1} = aa^{-1} = e.$$

Similarly, $(b^{-1}a^{-1})(ab) = e$. Hence, by definition, $(ab)^{-1} = b^{-1}a^{-1}$. $\square$

We promised to list a piece of the argument given above as a separate lemma. We keep this promise and write

Lemma 2.2.2.   In any group G, if $a, b, c \in G$ and
   (a) $ab = ac$, then $b = c$.
   (b) $ba = ca$, then $b = c$.

Before leaving these results, note that if G is the group of real numbers under $+$, then Part (c) of Lemma 2.2.1 translates into the familiar $-(-a) = a$.

There is only a scant bit of mathematics in this section; accordingly, we give only a few problems. No indication is given as to the difficulty of these.

Problems

1. Suppose that G is a set closed under an associative operation such that
    1. Given $a, y \in G$, there is an $x \in G$ such that $ax = y$.
    2. Given $a, w \in G$, there is a $u \in G$ such that $ua = w$.
    Show that G is a group.

2. If G is a *finite* set closed under an associative operation such that $ax = ay$ forces $x = y$ and $ua = wa$ forces $u = w$, for every $a, x, y, u, w \in G$, prove that G is a group. (This is a repeat of a problem given earlier. It will be used in the body of the text later.)

3. If G is a group in which $(ab)^i = a^i b^i$ for three consecutive integers i, prove that G is abelian.

4. Show that the result of Problem 3 would not always be true if the word "three" were replaced by "two". In other words, show that there is a group G and consecutive numbers $i, i+1$ such that G is not abelian but does have the property that $(ab)^i = a^i b^i$ and $(ab)^{i+1} = a^{i+1} b^{i+1}$ for all a, b in G.

5. Let G be a group in which $(ab)^3 = a^3 b^3$ and $(ab)^5 = a^5 b^5$ for all $a, b \in G$. Show that G is abelian.

6. Let G be a group in which $(ab)^n = a^n b^n$ for some fixed integer $n > 1$ for all $a, b \in G$. For all $a, b \in G$, prove that:
    (a) $(ab)^{n-1} = b^{n-1} a^{n-1}$.
    (b) $a^n b^{n-1} = b^{n-1} a^n$.
    (c) $(aba^{-1}b^{-1})^{n(n-1)} = e$.
    [**Hint for Part (c):** Note that $(aba^{-1})^r = ab^r a^{-1}$ for all integers r.]

3. SUBGROUPS

In order for us to find out more about the makeup of a given group G, it may be too much of a task to tackle all of G head-on. It might be desirable to focus our attention on appropriate pieces of G, which are smaller, over which we have some control, and are such that the information gathered about them can be used to get relevant information and insight about G itself. The question then becomes: What should serve as suitable pieces for this kind of dissection of G? Clearly, whatever we choose as such pieces, we want them to reflect the fact that G is a group, not merely any old set.

A group is distinguished from an ordinary set by the fact that it is endowed with a well-behaved operation. It is thus natural to demand that such pieces above behave reasonably with respect to the operation of G. Once this is granted, we are led almost immediately to the concept of a subgroup of a group.

> **Definition.** A nonempty subset, H, of a group G is called a *subgroup* of G if, relative to the product in G, H itself forms a group.

We stress the phrase "relative to the product in G." Take, for instance, the subset $A = \{1, -1\}$ in Z, the set of integers. Under the multiplication of integers, A is a group. But A is *not* a subgroup of Z viewed as a group with respect to $+$.

Every group G automatically has two obvious subgroups, namely G itself and the subgroup consisting of the identity element, e, alone. These two subgroups we call *trivial subgroups*. Our interest will be in the remaining ones, the *proper subgroups* of G.

Before proceeding to a closer look at the general character of subgroups, we want to look at some specific subgroups of some particular, explicit groups. Some of the groups we consider are those we introduced as examples in Section 1; we maintain the numbering given there for them. In some of these examples we shall verify that certain specified subsets are indeed subgroups. We would strongly recommend that the readers carry out such a verification in lots of the others and try to find other examples for themselves.

In trying to verify whether or not a given subset of a group is a subgroup, we are spared checking one of the axioms defining a group, namely the associative law. Since the associative law holds universally in a group G, given any subset A of G and any three elements of A, then the associative law certainly holds for them. So we must check, for a given subset A of G, whether A is closed under the operation of G, whether e is in A, and finally, given $a \in A$, whether a^{-1} is also in A.

Note that we can save one more calculation. Suppose that $A \subset G$ is nonempty and that given $a, b \in A$, then $ab \in A$; suppose further that given $a \in A$, then $a^{-1} \in A$. Then we assert that $e \in A$. For pick

$a \in A$; then $a^{-1} \in A$ by supposition, hence $aa^{-1} \in A$, again by supposition. Since $aa^{-1} = e$, we have that $e \in A$. Thus a is a subgroup of G. In other words,

Lemma 2.3.1. A nonempty subset $A \subset G$ is a subgroup of G if and only if A is closed with respect to the operation of G and, given $a \in A$, then $a^{-1} \in A$.

We now consider some examples.

EXAMPLES

1. Let G be the group $\mathbb{Z}$ of integers under $+$ and let H be the set of even integers. We claim that H is a subgroup of $\mathbb{Z}$. Why? Is H closed, that is, given $a, b \in H$, is $a + b \in H$? In other words, if a, b are even integers, is $a + b$ an even integer? The answer is yes, so H is certainly closed under $+$. Now to the inverse. Since the operation in $\mathbb{Z}$ is $+$, the inverse of $a \in \mathbb{Z}$ relative to this operation is $-a$. If $a \in H$, that is, if a is even, then $-a$ is also even, hence $-a \in H$. In short, H is a subgroup of $\mathbb{Z}$ under $+$.

2. Let G once again be the group $\mathbb{Z}$ of integers under $+$. In Example 1, H, the set of even integers, can be described in another way: namely H consists of all multiples of 2. There is nothing particular in Example 1 that makes use of 2 itself. Let $m > 1$ be any integer and let H_m consist of all multiples of m in Z. We leave it to the reader to verify that H_m is a subgroup of Z under $+$.

3. Let S be any nonempty set and let $G = A(S)$. If $a \in S$, let $H(a) = \{ f \in A(S) | f(a) = a \}$. We claim that $H(a)$ is a subgroup of G. For if $f, g \in H(a)$, then $(fg)(a) = f(g(a)) = f(a) = a$, since $f(a) = g(a) = a$; thus $fg \in H(a)$. Also, if $f \in H(a)$, then $f(a) = a$, so that $f^{-1}(f(a)) = f^{-1}(a)$; but $f^{-1}(f(a)) = f^{-1}(a) = i(a) = a$. Thus, since $a = f^{-1}(f(a)) = f^{-1}(a)$, we have that $f^{-1} \in H(a)$. Consequently, $H(a)$ is a subgroup of G.

4. Let G be Example 6 in Section 1, and H, Example 7. Then H is a subgroup of G (see Problem 3 in Section 1).

5. Let G be as in Example 6, and K as in Example 8 in Section 1. Then $K \subset H \subset G$ and K is a subgroup of both H and of G.

6. Let $\mathbf{C}'$ be the nonzero complex numbers as a group under the multiplication of complex numbers. Let $V = \{a \in \mathbf{C}' | |a| \text{ is rational}\}$. Then V is a subgroup of $\mathbf{C}'$. For if $|a|$ and $|b|$ are rational, then $|ab| = |a| |b|$ is rational, so $ab \in V$; also, $|a^{-1}| = 1/|a|$ is rational, hence $a^{-1} \in V$. Therefore, V is a subgroup of $\mathbf{C}'$.

7. Let $\mathbf{C}'$ and V be as above and let

$$U = \{a \in \mathbf{C}' | a = \cos\theta + i\sin\theta, \ \theta \text{ any real}\}.$$

If $a = \cos\theta + i\sin\theta$ and $b = \cos\psi + i\sin\psi$, we saw in Chapter 1 that $ab = \cos(\theta + \psi) + i\sin(\theta + \psi)$, so that $ab \in U$, and that $a^{-1} = \cos\theta - i\sin\theta = \cos(-\theta) + i\sin(-\theta) \in U$. Also, $|a| = 1$, since $|a| = \sqrt{\cos^2\theta + \sin^2\theta} = 1$. Therefore, $U \subset V \subset \mathbf{C}'$ and U is a subgroup both of V and of $\mathbf{C}'$.

8. Let $\mathbf{C}', U, V$ be as above, and let $n > 1$ be an integer. Let $\theta_n = \cos(2\pi/n) + i\sin(2\pi/n)$, and let $B = \{1, \theta_n, \theta_n^2, \ldots, \theta_n^{n-1}\}$. Since $\theta^n = 1$ (as we saw by De Moivre's Theorem), it is easily checked that B is a subgroup of U, V, and $\mathbf{C}'$, and is of order n.

9. Let G be any group and let $a \in G$. The set $A = \{a^i | i \text{ any integer}\}$ is a subgroup of G. For, by the rules of exponents, if $a^i \in A$ and $a^j \in A$, then $a^i a^j = a^{i+j}$, so is in A. Also, $(a^i)^{-1} = a^{-i}$, so $(a^i)^{-1} \in A$. This makes A into a subgroup of G.

A *is the cyclic subgroup of G generated by a in the following sense.*

Definition. The *cyclic subgroup of G generated by a* is the set $\{a^i | i \text{ any integer}\}$. It is denoted (a).

Note that if e is the identity element of G, then $(e) = \{e\}$. In Example 8, the group is the cyclic group (θ_n) of $\mathbf{C}$ generated by θ_n.

10. Let G be any group; for $a \in G$ let $C(a) = \{g \in G | ga = ag\}$. We claim that $C(a)$ is a subgroup of G. First, the closure of $C(a)$. If

$g, h \in C(a)$, then $ga = ag$ and $ha = ah$, thus $(gh)a = g(ha) = g(ah)$ $= (ga)h = (ag)h = a(gh)$ (by the repeated use of the associative law), hence $gh \in C(a)$. Also, if $g \in C(a)$, then from $ga = ag$ we have $g^{-1}(ga)g^{-1} = g^{-1}(ag)g^{-1}$, which simplifies to $ag^{-1} = g^{-1}a$; whence $g^{-1} \in C(a)$. $C(a)$ is thereby a subgroup of G.

These particular subgroups $C(a)$ will come up later for us and they are given a special name. We call $C(a)$ the *centralizer* of a in G. If in a group $ab = ba$, we say that a and b *commute*. Thus $C(a)$ is the set of all elements in G that commute with a.

11. Let G be any group and let $Z(G) = \{z \in G | zx = xz \text{ all } x \in G\}$. We leave it to the reader to verify that $Z(G)$ is a subgroup of G. It is called the *center* of G.

12. Let G be any group and H a subgroup of G. For $a \in G$, let $a^{-1}Ha = \{a^{-1}ha | h \in H\}$. We assert that $a^{-1}Ha$ is a subgroup of G. If $x = a^{-1}h_1a$ and $y = a^{-1}h_2a$, $h_1, h_2 \in H$, then $xy = (a^{-1}h_1a)(a^{-1}h_2a)$ $= a^{-1}(h_1h_2)a$ (associative law), and since H is a subgroup of G, $h_1h_2 \in H$; therefore, $a^{-1}(h_1h_2)a \in a^{-1}Ha$, which says that $xy \in a^{-1}Ha$. Thus $a^{-1}Ha$ is closed. Also, if $x = a^{-1}ha \in a^{-1}Ha$, then, as is easily verified, $x^{-1} = (a^{-1}ha)^{-1} = a^{-1}h^{-1}a \in a^{-1}Ha$. Therefore, $a^{-1}Ha$ is a subgroup of G.

An even dozen seems to be about the right number of examples, so we go on to other things. Lemma 2.3.1 points out for us what we need in order that a given subset of a group be a subgroup. In an important special case we can make a considerable saving in checking whether a given subset H is a subgroup of G. This is the case in which H is *finite*.

Lemma 2.3.2. Suppose that G is a group and H a nonempty *finite* subset of G closed under the product in G. Then H is a subgroup of G.

Proof. By Lemma 2.3.1 we must show that $a \in H$ implies $a^{-1} \in H$. If $a = e$, then $a^{-1} = e$ and we are done. Suppose then that $a \neq e$; consider the elements $a, a^2, \ldots, a^{n+1}$, where $n = |H|$, the order of H. *Here we have written down $n + 1$ elements, all of them in H, since H is closed, which has only n distinct elements.* How can this be? Only if some two of the elements listed are equal; put another way, only if $a^i = a^j$ for some $1 \leq i < j \leq n + 1$. But then, by the cancellation property in groups, $a^{j-i} = e$. Since $j - i \geq 1$, $a^{j-i} \in H$, hence $e \in H$. However, $j - i - 1 \geq 0$, so $a^{j-i-1} \in H$ and $aa^{j-i-1} = a^{j-i} = e$, whence $a^{-1} = a^{j-i-1} \in H$. This proves the lemma. $\square$

An immediate, but nevertheless important, corollary to Lemma 2.3.2 is the

Corollary. If G is a finite group and H a nonempty subset of G closed under multiplication, then H is a subgroup of G.

Problems

Easier Problems

1. If A, B are subgroups of G, show that $A \cap B$ is a subgroup of G.

2. What is the cyclic subgroup of Z generated by -1 under $+$?

3. Let S_3 be the symmetric group of degree 3. Find all the subgroups of S_3.

4. Verify that $Z(G)$, the center of G, is a subgroup of G. (See Example 11.)

5. If $C(a)$ is the centralizer of a in G (Example 10), prove that $Z(G) = \bigcap_{a \in G} C(a)$.

6. Show that $a \in Z(G)$ if and only if $C(a) = G$.

7. In S_3, find $C(a)$ for each $a \in S_3$.

8. If G is an abelian group and if $H = \{a \in G | a^2 = e\}$, show that H is a subgroup of G.

9. Give an example of a nonabelian group for which the H in Problem 8 is *not* a subgroup.

10. If G is an abelian group and $n > 1$ an integer, let $A_n = \{a^n | a \in G\}$. Prove that A_n is a subgroup of G.

*11. If G is an abelian group and $H = \{a \in G | a^{n(a)} = e$ for some $n(a) > 1$ depending on $a\}$, prove that H is a subgroup of G.

   We say that a group G is *cyclic* if there exists an $a \in G$ such that every $x \in G$ is a power of a, that is, $x = a^j$ for some j. We call a a *generator* of G in this case.

12. Prove that a cyclic group is abelian.

13. If G is cyclic, show that every subgroup of G is cyclic.

14. If G has no proper subgroups, prove that G is cyclic.

15. If G is a group and H a nonempty subset of G such that, given $a, b \in H$, then $ab^{-1} \in H$, prove that H is a subgroup of G.

Middle-Level Problems

16. If G has no proper subgroups, prove that G is cyclic of order p, where p is a prime number. (This sharpens the result of Problem 14.)

17. If G is a group and $a, x \in G$, prove that $C(x^{-1}ax) = x^{-1}C(a)x$. [See Examples 10 and 12 for the definitions of $C(b)$ and of $x^{-1}C(a)x$.]

18. If S is a nonempty set and $X \subset S$, show that $T(X) = \{f \in A(S) | f(X) \subset X\}$ is a subgroup of $A(S)$ if X is finite.

19. If A, B are subgroups of an abelian group G, let $AB = \{ab | a \in A, b \in B\}$. Prove that AB is a subgroup of G.

20. Give an example of a group G and two subgroups A, B of G such that AB is *not* a subgroup of G.

21. If A, B are subgroups of G such that $b^{-1}Ab \subset A$ for all $b \in B$, show that AB is a subgroup of G.

*22. If A and B are finite subgroups, of orders m and n, respectively, of the abelian group G, prove that AB is a subgroup of order mn *if m and n are relatively prime.*

23. What is the order of AB in Problem 22 if m and n are not relatively prime?

24. If H is a subgroup of G, let $N = \bigcap_{x \in G} x^{-1}Hx$. Prove that N is a subgroup of G such that $y^{-1}Ny = N$ for every $y \in G$.

Harder Problems

25. Let $S, X, T(X)$ be as in Problem 18 (but X no longer finite). Give an example of a set S and an infinite subset X such that $T(X)$ is *not* a subgroup of $A(S)$.

*26. Let G be a group, H a subgroup of G. Let $Hx = \{hx | h \in H\}$. Show that, given $a, b \in G$, then $Ha = Hb$ or $Ha \cap Hb = \varnothing$.

27. If in Problem 26 H is a finite subgroup of G, prove that Ha and Hb have the same number of elements. What is this number?

28. Let M, N be subgroups of G such that $x^{-1}Mx \subset M$ and $x^{-1}Nx \subset N$ for all $x \in G$. Prove that MN is a subgroup of G and that $x^{-1}(MN)x \subset MN$ for all $x \in G$.

*29. If M is a subgroup of G such that $x^{-1}Mx \subset M$ for all $x \in G$, prove that actually $x^{-1}Mx = M$.

30. If M, N are such that $x^{-1}Mx = M$ and $x^{-1}Nx = N$ for all $x \in G$ and if $M \cap N = (e)$, prove that $mn = nm$ for any $m \in M, n \in N$. (**Hint:** Consider the element $mnm^{-1}n^{-1}$.)

4.   LAGRANGE'S THEOREM

We are about to derive the first real group-theoretic result of importance. Although its proof is relatively easy, this theorem is like the A-B-C's for finite groups and has interesting implications in number theory.

As a matter of fact, those of you who solved Problems 16 and 17 of Section 3 have all the necessary ingredients to effect a proof of the result. The theorem simply states that in a finite group the order of a subgroup divides the order of the group.

To smooth the argument of this theorem—which is due to Lagrange —and for use many times later, we make a short detour into the realm of set theory.

Just as the concept of "function" runs throughout most phases of mathematics, so also does the concept of "relation". A *relation* is a statement $a R b$ about the elements $a, b \in S$. If S is the set of integers, $a = b$ is a relation on S. Similarly, $a < b$ is a relation on S, as is $a \leq b$.

Definition. A relation ~ on a set S is called an *equivalence relation* if its satisfies:

(a) $a \sim a$ (*reflexivity*)
(b) $a \sim b$ implies that $b \sim a$ (*symmetry*)
(c) $a \sim b$, $b \sim c$ implies that $a \sim c$ (*transitivity*)
for all $a, b, c \in S$.

Of course, equality, $=$, is an equivalence relation, so the general notion of equivalence relation is a generalization of that of equality. To be exact, an equivalence relation measures equality with regard to some attribute. This vague remark may become clearer after we see some examples.

EXAMPLES

1. Let S be all the items for sale in a grocery store; we declare $a \sim b$, for $a, b \in S$, if the price of a equals that of b. Clearly, the defining rules of an equivalence relation hold for this ~ . Note that in measuring this "generalized equality" on S we ignore all properties of the elements of S other than their price. So $a \sim b$ if they are equal as far as the attribute of price is concerned.

2. Let S be the integers and $n > 1$ a fixed integer. We define $a \sim b$ for $a, b \in S$ if $n|(a - b)$. We verify that this is an equivalence relation. Since $n|0 = a - a$, we have $a \sim a$. Because $n|(a - b)$ implies that

$$n|(b - a) = -(a - b),$$

we have that $a \sim b$ implies that $b \sim a$. Finally, if $a \sim b$ and $b \sim c$, then $n|(a - b)$ and $n|(b - c)$; hence $n|((a - b) + (b - c))$, that is, $n|(a - c)$. Therefore, $a \sim c$.

This relation on the integers is of great importance in number theory and is called *congruence modulo n*; when $a \sim b$ we write this as $a \equiv b \bmod(n)$ [or, sometimes, as $a \equiv b(n)$], which is read "a congruent to b mod n." We'll be running into it very often from now on. As we shall see, this is a special case of a much wider phenomenon in groups.

3. We generalize Example 2. Let G be a group and H a subgroup of G. For $a, b \in G$, define $a \sim b$ if $ab^{-1} \in H$. Since $e \in H$ and $e = aa^{-1}$,

we have that $a \sim a$. Also, if $ab^{-1} \in H$, then since H is a subgroup of G, $(ab^{-1})^{-1} \in H$. But $(ab^{-1})^{-1} = (b^{-1})^{-1}a^{-1} = ba^{-1}$, so $ba^{-1} \in H$, hence $b \sim a$. This tells us that $a \sim b$ implies that $b \sim a$. Finally, if $a \sim b$ and $b \sim c$, then $ab^{-1} \in H$ and $bc^{-1} \in H$. But $(ab^{-1})(bc^{-1}) = ac^{-1}$, whence $ac^{-1} \in H$ and therefore $a \sim c$. We have shown the transitivity of $\sim$, thus $\sim$ is an equivalence relation on G.

Note that if $G = \mathbb{Z}$, the group of integers under $+$, and H is the subgroup consisting of all multiplies of n, for $n > 1$ a fixed integer, then $ab^{-1} \in H$ translates into $a \equiv b(n)$. So congruence mod n is a very special case of the equivalence we have defined in Example 3.

It is this equivalence relation that we shall use in proving Lagrange's theorem.

4. Let G be any group. For $a, b \in G$ we declare that $a \sim b$ if there exists an $x \in G$ such that $b = x^{-1}ax$. We claim that this defines an equivalence relation on G. First, $a \sim a$ for $a = eae^{-1}$. Second, if $a \sim b$, then $b = x^{-1}ax$, hence $a = (x^{-1})^{-1}b(x^{-1})$, so that $b \sim a$. Finally, if $a \sim b$, $b \sim c$, then $b = x^{-1}ax$, $c = y^{-1}by$ for some $x, y \in G$. Thus $c = y^{-1}(x^{-1}ax)y = (xy)^{-1}a(xy)$, and so $a \sim c$. We have established that this defines an equivalence relation on G.

This relation, too, plays an important role in group theory and is given the special name *conjugacy*. When $a \sim b$ we say that "*a and b are conjugate in G*." Note that if G is abelian, then $a \sim b$ if and only if $a = b$.

We could go on and on to give numerous interesting examples of equivalence relations, but this would sidetrack us from our main goal in this section. There will be no lack of examples in the problems at the end of this section.

We go on with our discussion and make the

Definition. If $\sim$ is an equivalence relation on S, then $[a]$, the *class* of a, is defined by $[a] = \{b \in S | b \sim a\}$.

Let us see what the class of a is in the two examples, Examples 3 and 4, just given.

In Example 3, $a \sim b$ if $ab^{-1} \in H$, that is, if $ab^{-1} = h$, for some $h \in H$. Thus $a \sim b$ implies that $a = hb$. On the other hand, if $a = kb$ where

$k \in H$, then $ab^{-1} = (kb)b^{-1} = k \in H$, so $a \sim b$ if and only if $a \in Hb = \{hb|h \in H\}$. Therefore, $[b] = Hb$.

The set Hb is called a *right coset of H in G*. We ran into such in Problem 26 of Section 3. Note that $b \in Hb$, since $b = eb$ and $e \in H$ (also because $b E[b] = Hb$). Right cosets, and left handed counterparts of them called *left cosets*, play important roles in what follows.

In Example 4, we defined $a \sim b$ if $b = x^{-1}ax$ for some $x \in G$. Thus $[a] = \{x^{-1}ax|x \in G\}$. We shall denote $[a]$ in this case as cl(a) and call it the *conjugacy class* of a in G. If G is abelian, then cl(a) consists of a alone. In fact, if $a \in Z(G)$, the center of G, then cl(a) consists merely of a.

The notion of conjugacy and its properties will crop up again often, especially in Section 11.

We leave the examination of the class of an element a in Example 2 to later in this chapter.

The important influence that an equivalence relation has on a set is to break it up and *partition* it into nice disjoint pieces.

Theorem 2.4.1. If $\sim$ is an equivalence relation on S, then $S = \cup[a]$, where this union runs over one element from each class, and where $[a] \neq [b]$ implies that $[a] \cap [b] = \varnothing$. That is $\sim$ partitions S into equivalence classes.

Proof. Since $a \in [a]$, we have $\cup_{a \in S}[a] = S$. The proof of the second assertion is also quite easy. We show that if $[a] \neq [b]$, then $[a] \cap [b] = \varnothing$, or, what is equivalent to this, if $[a] \cap [b] \neq \varnothing$, then $[a] = [b]$.

Suppose, then, that $[a] \cap [b] \neq \varnothing$; let $c \in [a] \cap [b]$. By definition of class, $c \sim a$ since $c \in [a]$ and $c \sim b$ since $c \in [b]$. Therefore, $a \sim c$ by property 2) of $\sim$, and so, since $a \sim c$ and $c \sim b$, we have $a \sim b$. Thus $a \in [b]$; if $x \in [a]$, then $x \sim a, a \sim b$ gives us that $x \sim b$, hence $x \in [b]$. Thus $[a] \subset [b]$. The argument is obviously symmetric in a and b, so we have $[b] \subset [a]$, whence $[a] = [b]$, and our assertion above is proved.

The theorem is now completely proved. $\square$

We now can prove a famous result of Lagrange.

Theorem 2.4.2 (Lagrange's Theorem). If G is a finite group and H is a subgroup of G, then the order of H divides the order of G.

Proof. Let us look back at Example 3, where we established that the relation $a \sim b$ if $ab^{-1} \in H$ is an equivalence relation and that

$$[a] = Ha = \{ha|h \in H\}.$$

Let k be the number of distinct classes—call them $Ha_1, \ldots, Ha_k$. By Theorem 6.4.1, $G = Ha_1 \cup Ha_2 \cup \cdots \cup Ha_k$ and we know that $Ha_j \cap Ha_i = \varnothing$ if $i \neq j$.

We assert that any Ha_i has $|H|$ = order of H number of elements. Map $H \to Ha_i$ by sending $h \to ha_i$. We claim that this map is 1-1, for if $ha_i = h'a_i$, then by cancellation in G we would have $h = h'$; thus the map is 1-1. It is definitely onto by the very definition of Ha_i. So H and Ha_i have the same number, $|H|$, of elements.

Since $G = Ha_1 \cup \cdots \cup Ha_k$ and the Ha_i are disjoint and each Ha_i has $|H|$ elements, we have that $|G| = k|H|$. Thus $|H|$ divides $|G|$ and Lagrange's Theorem is proved. $\square$

> *Although Lagrange sounds like a French name, J. L. Lagrange (1736–1813) was actually Italian, having been born and brought up in Turin. He spent most of his life, however, in France. He was a very great mathematician who made fundamental contributions to all the areas of mathematics of his day.*

If G is finite, the number of right cosets of H in G, namely, $|G|/|H|$, is called the *index* of H in G and is written as $i_G(H)$.

Recall that a group G is said to be *cyclic* if there is an element $a \in G$ such that every element in G is a power of a.

Theorem 2.4.3. A group G of prime order is cyclic.

Proof. If H is a subgroup of G then, by invoking Lagrange's Theorem, $|H| \mid |G| = p$, p a prime, so $|H| = 1$ or p. So if $H \neq (e)$, then $H = G$. If $a \neq e \in G$, then the powers of a, $\{a^i\}$, form a subgroup of

G different from (e). So this subgroup is all of G. This says that any $x \in G$ is of the form $x = a^i$, hence G is cyclic by the definition of cyclic group. $\square$

If G is finite and $a \in G$, we saw earlier in the proof of Lemma 2.3.2 that $a^{n(a)} = e$ for some $n(a) \geq 1$, depending on a. We make the

Definition. If G is finite, then the *order* of a, written $o(a)$, is the *least positive integer m* such that $a^m = e$.

Suppose that $a \in G$ has order m. Consider the set $A = \{e, a, a^2, \ldots, a^{m-1}\}$; we claim that A is a subgroup of G (since $a^m = e$) and that the m elements listed in A are distinct. We leave the verification of these claims to the reader. Thus $|A| = m = o(a)$. Since $|A| \mid |G|$, we have

Theorem 2.4.4. If G is finite and $a \in G$, then $o(a) \mid |G|$.

If $a \in G$, where G is finite, we have, by Theorem 2.4.4, $|G| = k \cdot o(a)$. Thus

$$a^{|G|} = a^{k \cdot o(a)} = \left(a^{o(a)} \right)^k = e^k = e.$$

We have proved the

Theorem 2.4.5. If G is a finite group of order $n = |G|$, then $a^n = e$ for all $a \in G$.

When we apply this last result to certain special groups arising in number theory, we shall obtain some classical number-theoretic results due to Fermat and Euler.

Let $\mathbb{Z}$ be the integers and let $n > 1$ be a fixed integer. We go back to Example 2 of equivalence relations, where we defined $a \equiv b(n)$ (a congruent to $b \bmod n$) if $n \mid (a - b)$. The class of a, $[a]$, consists of all $a + nk$, where k runs through all the integers. We call it the *congruence class* of a.

Given any integer b then, by Theorem 1.5.1, $b = qn + r$, where $0 \leq r < n$, thus $[b] = [r]$. So the n classes $[0], [1], \ldots, [n - 1]$ give us all the congruence classes. We leave it to the reader to verify that they are distinct.

Let $\mathbb{Z}_n = \{[0], [1], \ldots, [n - 1]\}$. We shall introduce two operations, $+$ and $\cdot$, in $\mathbb{Z}_n$. Under $+$ $\mathbb{Z}_n$ will form an abelian group; under $\cdot$ $\mathbb{Z}_n$ will not form a group, but a certain piece of it will become a group.

How to define $[a] + [b]$? What is more natural than to define

$$[a] + [b] = [a + b].$$

But there is a fly in the ointment. Is this operation $+$ in $\mathbb{Z}_n$ well defined? What does that mean? We can represent $[a]$ by many a's—for instance, if $n = 3$, $[1] = [4] = [-2] = \cdots$, yet we are using a *particular* a to define the addition. What we must show is that if $[a] = [a']$ and $[b] = [b']$, then $[a + b] = [a' + b']$, for then we will have $[a] + [b] = [a + b] = [a' + b'] = [a'] + [b']$.

Suppose that $[a] = [a']$; then $n|(a - a')$. Also from $[b] = [b']$, $n|(b - b')$, hence $n|((a - a') + (b - b')) = ((a + b) - (a' + b'))$. Therefore, $a + b \equiv a' + b'(n)$, and so $[a + b] = [a' + b']$.

So we now have a well-defined addition in $\mathbb{Z}_n$. The element $[0]$ acts as the identity element and $[-a]$ acts as $-[a]$, the inverse of $[a]$. We leave it to the reader to check out that $\mathbb{Z}_n$ is a group under $+$. It is a cyclic group of order n generated by $[1]$.

We summarize this all as

Theorem 2.4.6. $\mathbb{Z}_n$ forms a cyclic group under the addition $[a] + [b] = [a + b]$.

Having disposed of the addition in $\mathbb{Z}_n$, we turn to the introduction of a multiplication. Again, what is more natural than defining

$$[a] \cdot [b] = [ab]?$$

So, for instance, if $n = 9$, $[2][7] = [14] = [5]$, and $[3][6] = [18] = [0]$. Under this multiplication—we leave the fact that it is well defined to the reader—$\mathbb{Z}_n$ *does not form* a group. Since $[0][a] = [0]$ for all a, and the unit element under multiplication is [1], [0] cannot have a multiplicative inverse. Okay, why not try the nonzero elements $[a] \neq [0]$ as a candidate for a group under this product? Here again it is no go *if n is not a prime.* For instance, if $n = 6$, then $[2] \neq [0]$, $[3] \neq [0]$, yet $[2][3] = [6] = [0]$, so the nonzero elements do not, in general, give us a group.

So we ask: Can we find an appropriate piece of $\mathbb{Z}_n$ that will form a group under multiplication? Yes! Let $U_n = \{[a] \in \mathbb{Z}_n | (a, n) = 1\}$, in other words, $\{[a]$ where a is relatively prime to $n\}$. By the Corollary to Theorem 1.5.5, if $(a, n) = 1$ and $(b, n) = 1$, then $(ab, n) = 1$, so $[a][b] = [ab]$ yields that if $[a], [b] \in U_n$, then $[ab] \in U_n$. So U_n is closed. Associativity is easily checked, following from the associativity of the integers under multiplication. The identity element is easy to find, namely [1]. Multiplication is commutative in U_n.

Note that if $[a][b] = [a][c]$ where $[a] \in U_n$, then we have $[ab] = [ac]$, and so $[ab - ac] = [0]$. This says that $n | a(b - c) = ab - ac$; but a is relatively prime to n. By Theorem 1.5.5 one must have that $n | (b - c)$, and so $[b] = [c]$. In other words, we have the cancellation property in U_n. By Problem 2 of Section 2, U_n is a group.

What is the order of U_n? By the definition of U_n, $|U_n| =$ number of integers $1 \leq m < n$ such that $(m, n) = 1$. This number comes up often and we give it a name.

Definition. The *Euler φ-function,* $\varphi(n)$, is defined by $\varphi(1) = 1$ and, for $n > 1$, $\varphi(n) =$ the number of positive integers m with $1 \leq m < n$ such that $(m, n) = 1$.

Thus $|U_n| = \varphi(n)$. If $n = p$, a prime, we have $\varphi(p) = p - 1$. We see that $\varphi(8) = 4$ for only 1, 3, 5, 7 are less than 8 and positive and relatively prime to 8. We try another one, $\varphi(15)$; of the numbers $1 \leq m < 15$ relatively prime to 15 there are 1, 2, 4, 7, 8, 11, 13, 14, so $\varphi(15) = 8$.

Let us look at some examples of U_n.

1. $U_8 = \{[1], [3], [5], [7])\}$. Note that $[3][5] = [15] = [7]$, $[5]^2 = [25] = [1]$. In fact, U_8 is a group of order 4 in which $a^2 = e$ for every $a \in U_8$.

2. $U_{15} = \{[1], [2], [4], [7], [8], [11], [13], [14]\}$. Note that $[11][13] = [143] = [8]$, $[2]^4 = [1]$, and so on.

The reader should verify that $a^4 = e = [1]$ for every $a \in U_{15}$.

3. $U_9 = \{[1], [2], [4], [5], [7], [8]\}$. Note that $[2]^1 = [2]$, $[2]^2 = [4]$, $[2]^3 = [8]$, $[2^4] = [16] = [7]$, $[2]^5 = [32] = [5]$; also $[2]^6 = [2][2]^5 = [2][5] = [10] = [1]$. So the powers of $[2]$ give us every element in U_9. Thus U_9 is a *cyclic* group of order 6. What other elements in U_9 generate U_9?

In parallel to Theorem 2.4.6 we have

> **Theorem 2.4.7.** U_n forms an abelian group, under the product $[a][b] = [ab]$, of order $\varphi(n)$, where $\varphi(n)$ is the Euler φ-function.

An immediate consequence of Theorems 2.4.7 and 2.4.5 is a famous result in number theory.

> **Theorem 2.4.8 (Euler).** If a is an integer relatively prime to n, then $a^{\varphi(n)} \equiv 1 \bmod n$.

Proof. U_n forms a group of order $\varphi(n)$, so by Theorem 2.4.5, $g^{\varphi(n)} = e$ for all $g \in U_n$. This translates into $[a^{\varphi(n)}] = [a]^{\varphi(n)} = [1]$, which in turn translates into $n | (a^{\varphi(n)} - 1)$. Precisely, this says that $a^{\varphi(n)} \equiv 1(n)$. $\square$

A special case, where $n = p$ is a prime, is due to Fermat.

> **Corollary (Fermat).** If p is a prime and $p \nmid a$, then
>
> $$a^{p-1} \equiv 1 \quad (p).$$
>
> For any integer b, $b^p \equiv b \bmod p$.

Proof. Since $\varphi(p) = p - 1$, if $(a, p) = 1$, we have, by Theorem 2.4.8, that $a^{p-1} \equiv 1(p)$, hence $a^1 \cdot a^{p-1} \equiv a(p)$, so that $a^p \equiv a(p)$. If $p|b$, then $b \equiv 0(p)$ and $b^p \equiv 0(p)$, so that $b^p \equiv b(p)$. $\square$

Leonhard Euler (1707–1785) was probably the greatest scientist that Switzerland has produced. He was the most prolific of all mathematicians ever.

Pierre Fermat (1601–1665) was a great number theorist. Fermat's Last Theorem—which is not a theorem but a conjecture—asks if $a^n + b^n = c^n$, a, b, c, n integers, has only the trivial solution where $a = 0$ or $b = 0$ or $c = 0$ if $n > 2$.

One final cautionary word about Lagrange's Theorem. Its *converse* in general is *not* true. That is, if G is a finite group of order n, then it need not be true that for every divisor m of n there is a subgroup of G of order m. A group with this property is very special indeed, and its structure can be spelled out quite well and precisely.

Problems

Easier Problems

1. Verify that the relation $\sim$ is an equivalence relation on the set S given.
    (a) $S = \mathbb{R}$ reals, $a \sim b$ if $a - b$ is rational.
    (b) $S = \mathbb{C}$, the complex numbers, $a \sim b$ if $|a| = |b|$.
    (c) $S =$ straight lines in the plane, $a \sim b$ if a, b are parallel.
    (d) $S =$ set of all people, $a \sim b$ if they have the same color eyes.

2. The relation $\sim$ on the real numbers $\mathbb{R}$ defined by $a \sim b$ if both $a > b$ and $b > a$ is *not* an equivalence relation. Why not? What properties of an equivalence relation does it satisfy?

3. Let $\sim$ be a relation on a set S that satisfies (1) $a \sim b$ implies that $b \sim a$ and (2) $a \sim b$, $b \sim c$. These seem to imply that $a \sim a$. For if $a \sim b$, then by (1), $b \sim a$, so $a \sim b$, $b \sim a$, so by (2), $a \sim a$. If this argument is correct, then the relation $\sim$ must be an equivalence relation. Problem 2 shows that this is not so. What is wrong with the argument we have given?

4. Let S be a set, $\{S_\alpha\}$ nonempty subsets such that $S = \cup_\alpha S_\alpha$ and $S_\alpha \cap S_\beta = \varnothing$ if $\alpha \neq \beta$. Define an equivalence relation on S in such a way that the S_α are precisely all the equivalence classes.

5. Let G be a group and H a subgroup of G. Define, for $a, b \in G$, $a \sim b$ if $a^{-1}b \in H$. Prove that this defines an equivalence relation on G, and show that $[a] = aH = \{ah | h \in H\}$. The sets aH are called left *cosets* of H in G.

6. If G is S_3 and $H = \{i, f\}$, where $f : S \to S$ is defined by $f(x_1) = x_2$, $f(x_2) = x_1$, $f(x_3) = x_3$, list all the right cosets of H in G and list all the left cosets of H in G.

7. In Problem 6, is every *right coset of H* in G also a left coset of H in G?

8. If every right coset of H in G is a left coset of H in G, prove that $aHa^{-1} = H$ for all $a \in G$.

9. In $\mathbb{Z}_{16}$, write down all the cosets of the subgroup $H = \{[0], [4], [8], [12]\}$. (Since the operation in $\mathbb{Z}_n$ is $+$, write your coset as $[a] + H$.) (We don't need to distinguish between right cosets and left cosets, since $\mathbb{Z}_n$ is abelian under $+$.)

10. In Problem 9, what is $i_{\mathbb{Z}_{16}}(H)$? (Recall that we defined the index $i_G(H)$ as the number of right cosets in G.)

11. For any finite group G, show that there are as many distinct left cosets of H in G as there are right cosets of H in G.

12. If aH and bH are distinct left cosets of H in G, are Ha and Hb distinct right cosets of H in G? Prove that this is true or give a counterexample.

13. Find the orders of all the elements of U_{18}. Is U_{18} cyclic?

14. Find the orders of all the elements of U_{20}. Is U_{20} cyclic?

15. If p is a prime, show that the only solutions of $x^2 \equiv 1(p)$ are $x \equiv 1(p)$ or $x \equiv -1(p)$.

16. If G is a finite abelian group and $a_1, \ldots, a_n$ are all its elements, show that $x = a_1 a_2 \cdots a_n$ must satisfy $x^2 = e$.

17. If G is of odd order, what can you say about the x in Problem 16?

18. Using the results of Problems 15 and 16, prove that if p is an odd prime number, then $(p - 1)! \equiv -1(p)$. (This is known as *Wilson's Theorem*.) It is, of course, also true if $p = 2$.

19. In S_3, find the conjugates of all its elements.

20. In the group G of Example 4 of Section 1, find the conjugacy class of the element $T_{a,b}$. Describe it in terms of a and b.

21. Let G be the dihedral group of order 8 (see Example 10, Section 1). Find the conjugacy classes in G.

22. Verify Euler's Theorem for $n = 14$ and $a = 3$, and $a = 5$.

23. In U_{41}, show that there is an element a such that $[a]^2 = [-1]$, that is, an integer a such that $a^2 \equiv -1 \ (41)$.

24. If p is a prime number of the form $4n + 3$, show that we *cannot* solve

$$x^2 \equiv -1(p).$$

[**Hint:** Use Fermat's Theorem that $a^{p-1} \equiv 1 \ (p)$ if $p \nmid a$.]

25. Show that the nonzero elements in $\mathbb{Z}_n$ form a group under the product $[a][b] = [ab]$ if and only if n is a prime.

Middle-Level Problems

26. Let G be a group, H a subgroup of G, and let S be the set of all distinct right cosets of H in G, T the set of all left cosets of H in G. Prove that there is a 1-1 mapping of S onto T. (**Note:** The obvious map that comes to mind, which sends Ha into aH, is not the right one. See Problems 5 and 12.)

27. If $aH = bH$ forces $Ha = Hb$ in G, show that $aHa^{-1} = H$ for every $a \in G$.

28. If G is a cyclic group of order n, show that there are $\varphi(n)$ generators for G. Give their form explicitly.

29. If in a group G, $aba^{-1} = b^i$, show that $a^r b a^{-r} = b^{i^r}$ for all positive integers r.

30. If in G $a^5 = e$ and $aba^{-1} = b^2$, find $o(b)$ if $b \neq e$.

31. If $o(a) = m$ and $a^s = e$, prove that $m|s$.

32. Let G be a finite group, H a subgroup of G. Let $f(a)$ be the least positive m such that $a^m \in H$. Prove that $f(a)|o(a)$.

33. If $i \neq f \in A(S)$ is such that $f^p = e$, p a prime, and if for some $s \in S$, $f^j(s) = s$ for some $1 \leq j < p$, show that $f(s) = s$.

34. If $f \in A(S)$ has order p, p a prime, show that for every $s \in S$ the orbit of s under f has one or p elements. [**Recall:** The orbit of s under f is $\{f^j(s)|j$ any integer$\}$.]

35. If $f \in A(S)$ has order p, p a prime, and S is a finite set having n elements, where $(n, p) = 1$, show that for some $s \in S$, $f(s) = s$.

Harder Problems

36. If $a > 1$ is an integer, show that $n|\varphi(a^n - 1)$, where φ is the Euler φ-function. [**Hint:** Consider the integers $\bmod(a^n - 1)$.]

37. In a cyclic group of order n, show that for each integer m that divides n (including $m = 1$ and $m = n$) there are $\varphi(m)$ elements of order m.

38. Using the result of Problem 37, show that $n = \sum_{m|n}\varphi(m)$.

39. Let G be a finite abelian group of order n for which the number of solutions of $x^m = e$ is at most m for any m dividing n. Prove that G must be cyclic. [**Hint:** Let $\psi(m)$ be the number of elements in G of *order* m. Show that $\psi(m) \le \varphi(m)$ and use Problem 38.]

40. Using the result of Problem 39, show that U_p, if p is a prime, is cyclic. (This is a famous result in number theory; it asserts the existence of a *primitive root* mod p.)

41. Using the result of Problem 40, show that if p is a prime of the form $p = 4n + 1$, then we can solve $x^2 \equiv -1(p)$ (with x an integer).

42. Using Wilson's Theorem (see Problem 28), show that if p is a prime of the form $p = 4n + 1$ and if

$$y = 1 \cdot 2 \cdot 3 \cdots \frac{p-1}{2} = \left(\frac{p-1}{2}\right)!,$$

then $y^2 \equiv -1(p)$. (This gives another proof to the result in Problem 41.)

43. Let G be an abelian group of order n, $a_1, \ldots, a_n$ its element. Let $x = a_1 a_2 \cdots a_n$. Show that:

(a) If G has exactly one element $b \neq e$ such that $b^2 = e$, then $x = b$.

(b) If G has more than one element $b \neq e$ such that $b^2 = e$, then $x = e$.

(c) If n is *odd*, then $x = e$ (see Problem 16).

5. HOMOMORPHISMS AND NORMAL SUBGROUPS

In a certain sense the subject of group theory is built up out of three basic concepts: that of a homomorphism, that of a normal subgroup, and that of the factor or quotient group of a group by a normal subgroup. We discuss the first two of these in this section, and the third in Section 6.

Without further ado we introduce the first of these.

> **Definition.** Let G, G' be two groups; then the mapping $\varphi: G \to G'$ is a *homomorphism* if $\varphi(ab) = \varphi(a)\varphi(b)$ for all $a, b \in G$.
>
> (*Note*: This φ has nothing to do with the Euler φ-function.)

In this definition the product on the left side—in $\varphi(ab)$—is that of G, while the product $\varphi(a)\varphi(b)$ is that of G'. A short description of a homomorphism is that it *preserves* the operation of G. We do *not* insist that φ be onto; if it is, we'll say that it is. Before working out some facts about homomorphisms, we present some examples.

EXAMPLES

1. Let G be the group of all positive reals under the multiplication of reals, and G' the group of all reals under addition. Let $\varphi: G \to G'$ be defined by $\varphi(x) = \log_{10}x$ for $x \in G$. Since $\log_{10}(xy) = \log_{10}x + \log_{10}y$, we have $\varphi(xy) = \varphi(x) + \varphi(y)$, so φ is a homomorphism. It also happens to be onto and 1-1.

2. Let G be an *abelian* group and let $\varphi: G \to G'$ be defined by $\varphi(a) = a^2$. Since $\varphi(ab) = (ab)^2 = a^2b^2 = \varphi(a)\varphi(b)$, φ is a homomorphism of G into itself. It need not be onto; the reader should check that in U_8 (see Section 4) $a^2 = e$ for all $a \in U_8$, so $\{\varphi(a)\} = (e)$.

3. The example of U_8 above suggests the so-called *trivial homomorphism*. Let G be any group and G' any other; define $\varphi(x) = e'$, the unit element of G', for all $x \in G$. Trivially, φ is a homomorphism of G into G'. It certainly is not a very interesting one.

Another homomorphism always present is the identity mapping, i, of any group G into itself. Since $i(x) = x$ for all $x \in G$, clearly $i(xy) = xy = i(x)i(y)$. The map i is 1-1 and onto, but, again, is not too interesting as a homomorphism.

4. Let G be the group of integers under $+$ and $G' = \{1, -1\}$ the subgroup of the reals under multiplication. Define $\varphi(m) = 1$ if m is even, $\varphi(m) = -1$ if m is odd. The statement that φ is a homomorphism is merely a restatement of:

even + even = even, even + odd = odd, and odd + odd = even.

5. Let G be the group of all nonzero complex numbers under multiplication and let G' be the group of positive reals under multiplication. Let $\varphi : G \to G'$ be defined by $\varphi(a) = |a|$; then $\varphi(ab) = |ab| = |a|\,|b| = \varphi(a)\varphi(b)$, so φ is a homomorphism of G into G'. In fact, φ is onto.

6. Let G be the group in Example 6 of Section 1, and G' the group of nonzero reals under multiplication. Define $\varphi : G \to G'$ by $\varphi(T_{a,b}) = a$. That φ is a homomorphism follows from the product rule in G, namely $T_{a,b}T_{c,d} = T_{ac,ad+b}$.

7. Let $G = \mathbb{Z}$ be the group of integers under $+$ and let $G' = \mathbb{Z}_n$. Define $\varphi : G \to \mathbb{Z}_n$ by $\varphi(m) = [m]$. Since the addition in $\mathbb{Z}_n$ is defined by $[m] + [r] = [m + r]$, we see that $\varphi(m + r) = \varphi(m) + \varphi(r)$, so φ is indeed a homomorphism of $\mathbb{Z}$ onto $\mathbb{Z}_n$.

8. The following general construction gives rise to a well-known theorem. Let G be any group, and let $A(G)$ be the set of all 1-1 mappings of G onto itself—here we are viewing G merely as a set, forgetting about its multiplication. Define $T_a : G \to G$ by $T_a(x) = ax$ for every $x \in G$. *What is the product*, T_aT_b, *of T_a and T_b as mappings on G?* Well,

$$(T_aT_b)(x) = T_a(T_bx) = T_a(bx) = a(bx) = (ab)x = T_{ab}(x)$$

(we used the associative law). So we see that $T_aT_b = T_{ab}$.

Define the mapping $\varphi: G \rightarrow A(G)$ by $\varphi(a) = T_a$, for $a \in G$. The product rule for the T's translates into $\varphi(ab) = T_{ab} = T_a T_b = \varphi(a)\varphi(b)$, so φ is a homomorphism of G into $A(G)$. We claim that φ is 1-1; suppose that $\varphi(a) = \varphi(b)$, that is, $T_a = T_b$. Therefore, $a = T_a(e) = T_b(e) = b$, so φ is indeed 1-1. It is not onto in general—for instance, if G has order $n > 2$, then $A(G)$ has order $n!$, and since $n! > n$, φ doesn't have a ghost of a chance of being onto. It is easy to verify that the image of φ, $\varphi(G) = \{T_a | a \in G\}$, is a subgroup of $A(G)$.

The fact that φ is 1-1 suggests that perhaps 1-1 homomorphisms should play a special role. We single them out in the

> **Definition.**   The homomorphism $\varphi: G \rightarrow G'$ is called a *monomorphism* if φ is 1-1. A monomorphism that is onto is called an *isomorphism*. An isomorphism from G to G itself is called an *automorphism*.

One more definition.

> **Definition.**   The two groups G and G' are said to be *isomorphic* if there is a monomorphism of G onto G'.
>   We shall denote that G and G' are isomorphic by writing $G \simeq G'$.

This definition seems to be asymmetric, but, in point of fact, it is not. For if there is an isomorphism of G onto G', there is one of G' onto G (see Problem 2).

We shall discuss more thoroughly later what it means for two groups to be isomorphic. But now we summarize what we did in Example 8.

> **Theorem 2.5.1 (Cayley's Theorem).**   Every group G is isomorphic to some subgroup of $A(S)$, for an appropriate S.

The appropriate S we used was G itself. But there may be better choices. We shall see some in the problems to follow.

When G is finite, we can take the set S in Theorem 2.5.1 to be finite, in which case $A(S)$ is S_n and its elements are permutations. In this case, Cayley's Theorem is usually stated as:

A finite group can be represented as a group of permutations.

Arthur Cayley (1821–1895) was an English mathematician who worked in matrix theory, invariant theory, and many other parts of algebra.

This is a good place to discuss the importance of "isomorphism." Let φ be an isomorphism of G onto G'. We can view G' as a relabeling of G, using the label $\varphi(x)$ for the element x. Is this labeling consistent with the structure of G as a group? That is, if x is labeled $\varphi(x)$, y labeled $\varphi(y)$, what is xy labeled as? Since $\varphi(x)\varphi(y) = \varphi(xy)$, we see that xy is labeled as $\varphi(x)\varphi(y)$, so this renaming of the elements is consistent with the product in G. So two groups that are isomorphic—although they need not be equal—in a certain sense, as described above—are equal. Often, it is desirable to be able to identify a given group as isomorphic to some concrete group that we know.

We go on with more examples.

9. Let G be any group, $a \in G$ fixed in the discussion. Define $\varphi : G \to G$ by $\varphi(x) = a^{-1}xa$ for all $x \in G$. We claim that φ is an isomorphism of G onto itself. First,

$$\varphi(xy) = a^{-1}(xy)a = a^{-1}xa \cdot a^{-1}ya = \varphi(x)\varphi(y),$$

so φ is at least a homomorphism of G into itself. It is 1-1 for if $\varphi(x) = \varphi(y)$, then $a^{-1}xa = a^{-1}ya$, so by cancellation in G we get $x = y$. Finally, φ is onto, for $x = a^{-1}(axa^{-1})a = \varphi(axa^{-1})$ for any $x \in G$.

Here φ is called the *inner automorphism* of G *induced* by a. The notion of *automorphism* and some of its properties will come up in the problems.

One final example:

10. Let G be the group of reals under $+$ and let G' be the group of all nonzero complex numbers under multiplication. Define $\varphi : G \to G'$ by

$$\varphi(x) = \cos x + i \sin x.$$

We saw that $(\cos x + i \sin x)(\cos y + i \sin y) = \cos(x+y) + i \sin(x+y)$, hence $\varphi(x)\varphi(y) = \varphi(x+y)$ and φ is a homomorphism of G into G'. φ is not 1-1 because, for instance, $\varphi(0) = \varphi(2\pi) = 1$, nor is φ onto.

Now that we have a few examples in hand, we start a little investigation of homomorphisms. We begin with

> **Lemma 2.5.2.**   If φ is a homomorphism of G into G', then:
> (a) $\varphi(e) = e'$, the unit element of G'.
> (b) $\varphi(a^{-1}) = \varphi(a)^{-1}$ for all $a \in G$.

Proof.   Since $x = xe$, $\varphi(x) = \varphi(xe) = \varphi(x)\varphi(e)$; by cancellation in G' we get $\varphi(e) = e'$. Also, $\varphi(aa^{-1}) = \varphi(e) = e'$, hence $e' = \varphi(aa^{-1}) = \varphi(a)\varphi(a^{-1})$. $\square$

> **Definition.**   The *image* of φ, $\varphi(G)$, is $\varphi(G) = \{\varphi(a) | a \in G\}$.

We leave to the reader the proof of

> **Lemma 2.5.3.**   If φ is a homomorphism of G into G', then the image of φ is a subgroup of G'.

We singled out certain homomorphisms and called them monomorphisms. Their property was that they were 1-1. We want to measure how far a given homomorphism is from being a monomorphism. This prompts the

> **Definition.**   If φ is a homomorphism of G into G', then the *kernel* of φ, Ker (φ), is defined by $\text{Ker}(\varphi) = \{a \in G | \varphi(a) = e'\}$.

Ker (φ) measures the lack of 1-1'ness at one point e'. We claim that this lack is rather uniform. What is $W = \{x \in G | \varphi(x) = w'\}$ for a given $w' \in G'$? Clearly, if $k \in \text{Ker}(\varphi)$ and $\varphi(x) = w'$, then $\varphi(xk) = \varphi(x)\varphi(k) = \varphi(x)e' = w'$, so $xk \in W$. Also, if $\varphi(x) = \varphi(y) = w'$, then $\varphi(x) = \varphi(y)$, hence $\varphi(y)\varphi(x)^{-1} = e'$; but $\varphi(x)^{-1} = \varphi(x^{-1})$ by Lemma 2.5.2, so $e' = \varphi(y)\varphi(x)^{-1} = \varphi(y)\varphi(x^{-1}) = \varphi(yx^{-1})$, whence

$yx^{-1} \in \mathrm{Ker}\ (\varphi)$ and so $y \in \mathrm{Ker}\ (\varphi)\ x$. *Thus the inverse image of any element w' in* $\varphi(G) \subset G'$ *is the set* $\mathrm{Ker}\ (\varphi)\ x$, *where x is any* element in G such that $\varphi(x) = w'$.

We state this as

Lemma 2.5.4. If $w' \in G'$ is of the form $\varphi(x) = w'$, then $\{ y \in G | \varphi(y) = w'\} = \mathrm{Ker}\ (\varphi)\ x$.

We now shall study some basic properties of the kernels of homomorphisms.

Theorem 2.5.5. If φ is a homomorphism of G into G', then
(a) $\mathrm{Ker}\ (\varphi)$ is a subgroup of G.
(b) Given $a \in G$, $a^{-1}\ \mathrm{Ker}\ (\varphi)\ a \subset \mathrm{Ker}\ (\varphi)$.

Proof.   Although this is so important, its proof is easy. If $a, b \in \mathrm{Ker}\ (\varphi)$, then $\varphi(a) = \varphi(b) = e'$, hence $\varphi(ab) = \varphi(a)\varphi(b) = e'$, whence $ab \in \mathrm{Ker}\ (\varphi)$; so $\mathrm{Ker}\ (\varphi)$ is closed under product. Also $\varphi(a) = e'$ implies that $\varphi(a^{-1}) = \varphi(a^{-1}) = e'$, and so $a^{-1} \in \mathrm{Ker}\ (\varphi)$. Therefore, $\mathrm{Ker}\ (\varphi)$ is a subgroup of G. If $K \in \mathrm{Ker}\ (\varphi)$ and $a \in G$, then $\varphi(k) = e'$; consequently, $\varphi(a^{-1}ka) = \varphi(a^{-1})\varphi(k)\varphi(a) = \varphi(a^{-1})e'\varphi(a) = \varphi(a^{-1})\varphi(a) = \varphi(a^{-1}a) = \varphi(e) = e'$. This tells us that $a^{-1}ka \in \mathrm{Ker}\ (\varphi)$, hence $a^{-1}\mathrm{Ker}(\varphi)a \subset \mathrm{Ker}(\varphi)$. The theorem is now completely proved. $\square$

Corollary.   If φ is a homomorphism of G into G', then φ is a monomorphism if and only if $\mathrm{Ker}\ (\varphi) = (e)$.

Proof.   This result is really a corollary to Lemma 2.5.4. We leave the few details to the reader. $\square$

Property (b) of $\mathrm{Ker}\ (\varphi)$ in Theorem 2.5.5 is an interesting and basic one for a subgroup to enjoy. We ran into this property in the text material and problems earlier on several occasions. We use it to define the ultra-important class of subgroups of a group.

Definition.   The subgroup N of G is said to be a *normal subgroup* of G if $a^{-1}Na \subset N$ for every $a \in G$.

Of course, $\text{Ker}(\varphi)$, for any homomorphism, is a normal subgroup of G. As we shall see in the next section, every normal subgroup of G is the kernel of some appropriate homomorphism of G into an appropriate group G'. So in a certain sense the notions of homomorphism and normal subgroups will be shown to be equivalent.

Although we defined a normal subgroup via $a^{-1}Na \subset N$, we actually have $a^{-1}Na = N$. For if $a^{-1}Na \subset N$ for all $a \in G$, then $N = a(a^{-1}Na)a^{-1} \subset aNa^{-1} = (a^{-1})^{-1}Na^{-1} \subset N$. So $N = aNa^{-1}$ for every $a \in G$. Transposing, we have $Na = aN$; that is, every left coset of N in G is a right coset of N in G.

On the other hand, if every left coset of N in G is a right coset, given $a \in G$, then $Na = bN$ for some $b \in G$. But $a \in Na$, so $a \in bN$, hence $aN \subset (bN)N = b(NN) \subset bN$ ($NN \subset N$, since N is closed). Thus $aN \subset bN = Na$, and so $aNa^{-1} \subset N$, which is to say, N is normal in G.

We write "N is a normal subgroup of G" by the abbreviated symbol $N \triangleleft G$.

Note that $a^{-1}Na = N$ *does not* mean that $a^{-1}na = n$ for every $n \in N$. No—merely that the set of all $a^{-1}na$ is the same as the set of all n.

We have proved

Theorem 2.5.6. $N \triangleleft G$ if and only if every left coset of N in G is a right coset of N in G.

Before going any further, we pause to look at some examples of kernels of homomorphisms and normal subgroups.

EXAMPLES

1. If G is abelian, then every subgroup of G is normal, for $a^{-1}xa = x$ for every $a, x \in G$. The converse of this is *not* true. Nonabelian groups exist in which *every subgroup* is normal. See if you can find such an

example of order 8. Such nonabelian groups are called *Hamiltonian*, after the Irish mathematician W. R. Hamilton (1805–1865). The desired group of order 8 can be found in the *quaternions* of Hamilton, which we introduce in Chapter 4, Section 1.

In Example 1, $\varphi(x) = \log_{10}x$, and $\text{Ker}\,\varphi = \{x | \log_{10}x = 0\} = \{1\}$. In Example 2, where G is abelian, and $\varphi(x) = x^2$,

$$\text{Ker}(\varphi) = \{x \in G | x^2 = e\}.$$

The kernel of the trivial homomorphism of Example 3 is *all* of G. In Example 4, $\text{Ker}(\varphi)$ is the set of all even integers. In Example 5, $\text{Ker}(\varphi) = \{a \in \mathbf{C}' \,|\, |a| = 1\}$, which can be identified, from the polar form of a complex number, as $\text{Ker}(\varphi) = \{\cos x + i \sin x | x \text{ real}\}$. In Example 6, $\text{Ker}(\varphi) = \{T_{1,b} \in G | b \text{ real}\}$. In Example 7, $\text{Ker}(\varphi)$ is the set of all multiples of n. In Examples 8 and 9, the kernels consist of e alone, for the maps are monomorphisms. In Example 10, we see that $\text{Ker}(\varphi) = \{2\pi m | m \text{ any integer}\}$.

Of course, all the kernels above are normal subgroups of their respective groups. We should look at some normal subgroups, intrinsically in G itself, without recourse to the kernels of homomorphism. We go back to the examples of Section 1.

1. In Example 7, $H = \{T_{a,b} \in G | a \text{ rational}\}$. If $T_{x,y} \in G$, we leave it to the reader to check that $T_{x,y}^{-1}HT_{x,y} \subset H$ and so $H \triangleleft G$.

2. In Example 9 the subgroup $\{i, g, g^2, g^3\} \triangleleft G$. Here too we leave the checking to the reader.

3. In Example 10 the subgroup $H = \{i, h, h^2, \ldots, h^{n-1}\}$ is normal in G. This we also leave to the reader.

4. If G is any group, $Z(G)$, the *center* of G is a normal subgroup of G (see Example 11 of Section 3).

5. Let $G = S_3$; G has the elements i, f, g, g^2, fg, and gf, where $f(x_1) = x_2$, $f(x_2) = x_1$, $f(x_3) = x_3$ and $g(x_1) = x_2$, $g(x_2) = x_3$, $g(x_3) = x_1$. We claim that the subgroup $N = \{i, g, g^2\} \triangleleft S_3$. As we saw earlier (or can compute now), $fgf^{-1} = g^{-1} = g^2$, $fg^2f^{-1} = g$. $(fg)g(fg)^{-1} = fggg^{-1}f^{-1} = fgf^{-1} = g^2$, and so on. So $N \triangleleft S_3$ follows.

The material in this section has been a rather rich diet. It may not seem so, but the ideas presented, although simple, are quite subtle. We recommend that the reader digest the concepts and results thoroughly before going on. One way of seeing how complete this digestion is, is to take a stab at many of the almost infinite list of problems that follow. The material of this next section is even a richer diet, and even harder to digest. Avoid a mathematical stomachache later by assimilating this section well.

Problems

Easier Problems

1. Determine in each of the parts if the given mapping is a homomorphism. If so, identify its kernel and whether or not the mapping is 1-1 or onto.
   (a) $G = \mathbb{Z}$ under $+$, $G' = Z_n$, $\varphi(a) = [a]$ for $a \in \mathbb{Z}$.
   (b) G group, $\varphi: G \to G$ defined by $\varphi(a) = a^{-1}$ for $a \in G$.
   (c) G abelian group, $\varphi: G \to G$ defined by $\varphi(a) = a^{-1}$ for $a \in G$.
   (d) G group of all nonzero real numbers under multiplication, $G' = \{1, -1\}$, $\varphi(r) = 1$ if r is positive, $\varphi(r) = -1$ if r is negative.
   (e) G an abelian group, $n > 1$ a fixed integer, and $\varphi: G \to G$ defined by $\varphi(a) = a^n$ for $a \in G$.

2. Recall that $G \simeq G'$ means that G is isomorphic to G'. Prove that for all groups G_1, G_2, G_3:
   (a) $G_1 \simeq G_1$.
   (b) $G_1 \simeq G_2$ implies that $G_2 \simeq G_1$.
   (c) $G_1 \simeq G_2$, $G_2 \simeq G_3$ implies that $G_1 \simeq G_3$.

3. Let G be any group and $A(G)$ the set of all 1-1 mappings of G, as a set, onto itself. Define $L_a: G \to G$ by $L_a(x) = xa^{-1}$. Prove that:
   (a) $L_a \in A(G)$.
   (b) $L_a L_b = L_{ab}$.
   (c) The mapping $\psi: G \to A(G)$ defined by $\psi(a) = L_a$ is a monomorphism of G into $A(G)$.

4. In Problem 3 prove that for all $a, b \in G$, $T_a L_b = L_b T_a$, where T_a is defined as in Example 8.

5. In Problem 4, show that if $V \in A(G)$ is such that $T_a V = V T_a$ for all $a \in G$, then $V = L_b$ for some $b \in G$. (**Hint:** Acting on $e \in G$, find out what b should be.)

6. Prove that if $\varphi : G \to G'$ is a homomorphism, then $\varphi(G)$, the image of G, is a subgroup of G'.

7. Show that $\varphi : G \to G'$, where φ is a homomorphism, is a monomorphism if and only if Ker $(\varphi) = (e)$.

8. Find an isomorphism of G, the group of all real numbers under $+$, *onto* G', the group of all positive real numbers under multiplication.

9. Verify that if G is the group in Example 6 of Section 1, and $H = \{ T_{a,b} \in G | a \text{ rational} \}$, then $H \triangleleft G$.

10. Verify that in Example 9 of Section 1, the dihedral group of order 8, the set $H = \{ i, g, g^2, g^3 \}$ is a normal subgroup of G.

11. Verify that in Example 10 of Section 1, the subgroup $H = \{ i, h, h^2, \ldots, h^{n-1} \}$ is normal in G.

12. Prove that if $Z(G)$ is the center of G, then $Z(G) \triangleleft G$.

13. If G is a finite abelian group of order n and $\varphi : G \to G$ is defined by $\varphi(a) = a^m$ for all $a \in G$, find the necessary and sufficient condition that φ be an isomorphism of G onto itself.

14. If G is abelian and $\varphi : G \to G'$ is a homomorphism of G *onto* G', prove that G' is abelian.

15. If G is any group, $N \triangleleft G$, and $\varphi : G \to G'$ a homomorphism of G *onto* G', prove that the image, $\varphi(N)$, of N is a normal subgroup of G'.

16. If $N \triangleleft G$ and $M \triangleleft G$ and $MN = \{ mn | m \in M, n \in N \}$, prove that MN is a subgroup of G and that $MN \triangleleft G$.

17. If $M \triangleleft G$, $N \triangleleft G$, prove that $M \cap N \triangleleft G$.

18. If H is any subgroup of G and $N = \bigcap_{a \in G} a^{-1} H a$, prove that $N \triangleleft G$.

19. If H is a subgroup of G, let $N(H)$ be defined by the relation $N(H) = \{ a \in G | a^{-1} H a = H \}$. Prove that:
    (a) $N(H)$ is a subgroup of G, $N(H) \supset H$.

(b) $H \triangleleft N(H)$.

(c) If K is a subgroup of G such that $H \triangleleft K$, then $K \subset N(H)$. [So $N(H)$ is the largest subgroup of G in which H is normal.]

20. If $M \triangleleft G$, $N \triangleleft G$, and $M \cap N = (e)$, show that for $m \in M$, $n \in N$, $mn = nm$.

21. Let S be any set having more than two elements and $A(S)$ the set of all 1-1 mappings of S onto itself. If $s \in S$, we define $H(s) = \{ f \in A(S) | f(s) = s \}$. Prove that $H(s)$ *cannot* be a normal subgroup of $A(S)$.

22. Let $G = S_3$, the symmetric group of degree 3 and let $H = \{ i, f \}$, where $f(x_1) = x_2$, $f(x_2) = x_1$, $f(x_3) = x_3$.

(a) Write down all the left cosets of H in G.

(b) Write down all the right cosets of H in G.

(c) Is every left coset of H a right coset of H?

23. Let G be a group such that all subgroups of G are normal in G. If $a, b \in G$, prove that $ba = a^j b$ for some j.

24. If G_1, G_2 are two groups, let $G = G_1 \times G_2$, the Cartesian product of G_1, G_2 [i.e., G is the set of all ordered pairs (a, b) where $a \in G_1$, $b \in G_2$]. Define a product in G by $(a_1, b_1)(a_2, b_2) = (a_1 a_2, b_1 b_2)$.

(a) Prove that G is a group.

(b) Show that there is a monomorphism φ_1 of G_1 into G such that $\varphi_1(G_1) \triangleleft G$, given by $\varphi_1(a_1) = (a_1, e_2)$, where e_2 is the identity element of G_2.

(c) Do Part (b) also for G_2.

(d) Using the mappings φ_1, φ_2 of Parts (b) and (c), prove that $\varphi_1(G_1) \varphi_2(G_2) = G$ and $\varphi_1(G_1) \cap \varphi_2(G_2)$ is the identity element of G.

(e) Prove that $G_1 \times G_2 \simeq G_2 \times G_1$.

25. Let G be a group and let $W = G \times G$ as defined in Problem 24. Prove that:

(a) The mapping $\varphi : G \rightarrow W$ defined by $\varphi(a) = (a, a)$ is a monomorphism of G into W.

(b) The image $\varphi(G)$ in W [i.e., $\{(a, a) | a \in G \}$] is normal in W if and only if G is abelian.

Middle-Level Problems

*26. If G is a group and $a \in G$, define $\sigma_a : G \to G$ by $\sigma_a(g) = aga^{-1}$; we saw in Example 9 of this section that σ_a is an isomorphism of G onto itself, so $\sigma_a \in A(G)$, the group of all 1-1 mappings of G (as a set) onto itself. Define $\psi : G \to A(G)$ by $\psi(a) = \sigma_a$ for all $a \in G$. Prove that:
(a) ψ is a homomorphism of G into $A(G)$.
(b) $\text{Ker } \psi = Z(G)$, the center of G.

27. If θ is an automorphism of G and $N \triangleleft G$, prove that $\theta(N) \triangleleft G$.

28. Let θ, ψ be automorphisms of G, and let $\theta\psi$ be the product of θ and ψ as mappings on G. Prove that $\theta\psi$ is an automorphism of G, and that θ^{-1} is an automorphism of G, so that the set of all automorphisms of G is itself a group.

*29. A subgroup T of a group W is called *characteristic* if $\varphi(T) \subset T$ for *all automorphisms*, φ, of W. Prove that:
(a) M characteristic in G implies that $M \triangleleft G$.
(b) M, N characteristic in G implies that MN is characteristic in G.
(c) A normal subgroup of a group *need not* be characteristic. (This is quite hard; you must find an example of a group G and a noncharacteristic normal subgroup.)

30. Suppose that $|G| = pm$, where $p \nmid m$ and p is a prime. If H is a normal subgroup of order p in G, prove that H is characteristic.

31. Suppose that G is an abelian group of order $p^n m$ where $p \nmid m$ is a prime. If H is a subgroup of G of order p^n, prove that H is a characteristic subgroup of G.

32. Do Problem 31 even if G is not abelian if you happen to know that for some reason or other $H \triangleleft G$.

33. Suppose that $N \triangleleft G$ and $M \subset N$ is a characteristic subgroup of N. Prove that $M \triangleleft G$. (It is *not* true that if $M \triangleleft N$ and $N \triangleleft G$, then M must be normal in G. See Problem 50.)

34. Let G be a group, $\mathscr{A}(G)$ the group of all automorphisms of G. (See Problem 28.) Let $I(G) = \{\sigma_a | a \in G\}$, where σ_a is as defined in Problem 26. Prove that $I(G) \triangleleft \mathscr{A}(G)$.

35. Show that $Z(G)$, the center of G, is a characteristic subgroup of G.

36. If $N \triangleleft G$ and H is a subgroup of G, show that $H \cap N \triangleleft H$.

Harder Problems

37. If G is a nonabelian group of order 6, prove that $G \simeq S_3$.

38. Let G be a group and H a subgroup of G. Let $S = \{ Ha | a \in G \}$ be the set of all left cosets of H in G. Define, for $b \in G$, $T_b : S \rightarrow S$ by $T_b(Ha) = Hab^{-1}$.
   (a) Prove that $T_b T_c = T_{bc}$ for all $b, c \in G$ [therefore the mapping $\psi : G \rightarrow A(S)$ defined by $\psi(b) = T_b$ is a homomorphism].
   (b) Describe $\text{Ker}(\psi)$, the kernel of $\psi : G \rightarrow A(S)$.
   (c) Show that $\text{Ker}(\psi)$ is the largest normal subgroup of G lying in H [largest in the sense that if $N \triangleleft G$ and $N \subset H$, then $N \subset \text{Ker}(\psi)$].

39. Use the result of Problem 38 to redo Problem 37.

Recall that if H is a subgroup of G, then the *index* of H in G, $i_G(H)$, is the number of distinct left cosets of H in G (if this number is finite).

40. If G is a finite group, H a subgroup of G such that $n \nmid i_G(H)!$ where $n = |G|$, prove that there is a normal subgroup $N \neq (e)$ of G contained in H.

41. Suppose that you know that a group G of order 21 contains an element a of order 7. Prove that $A = (a)$, the subgroup generated by a, is normal in G. (**Hint:** Use the result of Problem 40.)

42. Suppose that you know that a group G of order 36 has a subgroup H of order 9. Prove that either $H \triangleleft G$ or there exists a subgroup $N \triangleleft G$, $N \subset H$, and $|N| = 3$.

43. Prove that a group of order 9 must be abelian.

44. Prove that a group of order p^2, p a prime, has a normal subgroup of order p.

45. Using the result of Problem 44, prove that a group of order p^2, p a prime, must be abelian.

46. Let G be a group of order 15; show that there is an element $a \neq e$ in G such that $a^3 = e$ and an element $b \neq e$ such that $b^5 = e$.

47. In Problem 46, show that both subgroups $A = \{e, a, a^2\}$ and $B = \{e, b, b^2, b^3, b^4\}$ are normal in G.

48. From the result of Problem 47, show that any group of order 15 is cyclic.

Very Hard Problems

49. Let G be a group, H a subgroup of G such that $i_G(H)$ is finite. Prove that there is a subgroup $N \subset H$, $N \triangleleft G$ such that $i_G(N)$ is finite.

50. Construct a group G such that G has a normal subgroup N, and N has a normal subgroup M (i.e., $N \triangleleft G$, $M \triangleleft N$), yet M is not normal in G.

51. Let G be a finite group, φ an automorphism of G such that φ^2 is the identity automorphism of G. Suppose that $\varphi(x) = x$ implies that $x = e$. Prove that G is abelian and $\varphi(a) = a^{-1}$ for all $a \in G$.

52. Let G be a finite group and φ an automorphism of G such that $\varphi(x) = x^{-1}$ for *more than three-fourths* of the elements of G. Prove that $\varphi(y) = y^{-1}$ for *all* $y \in G$, and so G is abelian.

6. FACTOR GROUPS

Let G be a group and N a normal subgroup of G. In proving Lagrange's Theorem we used, for an arbitrary subgroup H, the equivalence relations $a \sim b$ if $ab^{-1} \in H$. Let's try this out when N is normal and see if we can say a little more than one could say for just any old subgroup.

So, let $a \sim b$ if $ab^{-1} \in N$ and let $[a] = \{x \in G | x \sim a\}$. As we saw earlier, $[a] = Na$, the left coset of N in G containing a. Recall that in looking at Z_n we defined for it an operation $+$ via $[a] + [b] = [a + b]$. Why not try something similar for an arbitrary group G and a normal subgroup N of G?

So let $M = \{[a] | a \in G\}$, where $[a] = \{x \in G | xa^{-1} \in N\} = Na$. We define a product in M via $[a][b] = [ab]$. We shall soon show that M is a group under this product. *But first and foremost we must show that this product in M is well defined*. In other words, we must show that if

$[a] = [a']$ and $[b] = [b']$, then $[ab] = [a'b']$, for this would show that $[a][b] = [ab] = [a'b'] = [a'][b']$; equivalently, that *this product of classes does not depend on the particular representatives we use for the classes.*

Therefore let us suppose that $[a] = [a']$ and $[b] = [b']$. From the definition of our equivalence we have that $a' = na$, where $n \in N$. Similarly, $b' = mb$, where $m \in N$. Thus $a'b' = namb = n(ama^{-1})ab$; since $N \triangleleft G$, ama^{-1} is in N, so $n(ama^{-1})$ is also in N. So if we let $n_1 = n(ama^{-1})$, then $n_1 \in N$ and $a'b' = n_1ab$. But this tells us that $a'b' \in Nab$, so that $a'b' \sim ab$, from which we have that $[a'b'] = [ab]$, the exact thing we required to ensure that our product in M was well defined.

Thus M is now endowed with a well-defined product $[a][b] = [ab]$. We now verify the group axioms for M. Closure we have from the very definition of this product. If $[a]$, $[b]$, and $[c]$ are in M, then $[a]([b][c]) = [a][bc] = [a(bc)] = [(ab)c]$ (since the product in G is associative) $= [ab][c] = ([a][b])[c]$. Therefore, the associative law has been established for the product in M. What about a unit element? Why not try the obvious choice, namely $[e]$? We immediately see that $[a][e] = [ae] = [a]$ and $[e][a] = [ea] = [a]$, so $[e]$ does act as the unit element for M. Finally, what about inverses? Here, too, the obvious choice is the correct one. If $a \in G$, then $[a][a^{-1}] = [aa^{-1}] = [e]$, hence $[a^{-1}]$ acts as the inverse of $[a]$ relative to the product we have defined in M.

We want to give M a name, and better still, a symbol that indicates its dependence on G and N. The symbol we use for M is G/N (read "G over N or G mod N") and G/N is called the *factor group* or *quotient group* of G by N.

What we have shown is the very important

Theorem 2.6.1. If $N \triangleleft G$ and

$$G/N = \{[a] | a \in G\} = \{Na | a \in G\},$$

then G/N is a group relative to the operation $[a][b] = [ab]$.

One observation must immediately be made, namely

Theorem 2.6.2. If $N \triangleleft G$, then there is a homomorphism ψ of G onto G/N such that Ker (ψ), the kernel of ψ, is N.

Proof. The most natural mapping from G to G/N is the one that does the trick. Define $\psi : G \rightarrow G/N$ by $\psi(a) = [a]$. Our product as defined in G/N makes of ψ a homomorphism, for $\psi(ab) = [ab] = [a][b] = \psi(a)\psi(b)$. Since every element $X \in G/N$ is of the form $X = [b] = \psi(b)$ for some $b \in G$, ψ is onto. Finally, what is the kernel, Ker (ψ), of ψ? By definition, Ker $(\psi) = (a \in G | \psi(a) = E)$, where E is the unit element of G/N. But what is E? Nothing other than $E = [e] = Ne = N$, and $a \in$ Ker (ψ) if and only if $E = N = \psi(a) = Na$, But $Na = N$ tells us that is $a = ea \in Na = N'$, so we see that, Ker $(\psi) \subset N$. That $N \subset$ Ker (ψ)— which is easy—well leave to the reader. So Ker $(\psi) = N$. $\square$

Theorem 2.6.2 substantiates the remark we made in the preceding section that every normal subgroup N of G is the kernel of some homomorphism of G onto some group. The "some homomorphism" is the ψ defined above and the "some group" is G/N.

This construction of the factor group of G by N is possibly the single most important construction in group theory. In other algebraic systems we shall have analogous constructions, as we shall see later.

One might ask: Where in this whole affair did the normality of N in G enter? Why not do the same thing for any subgroup H of G? So let's try and see what happens. As before, we define

$$W = \{[a] | a \in G\} = \{Ha | a \in G\}$$

where the equivalence $a \sim b$ is defined by $ab^{-1} \in H$. We try to introduce a product in W as we did for G/N by defining $[a][b] = [ab]$. Is this product well defined? If $h \in H$, then $[hb] = [b]$, so for the product to be well defined, we would need that $[a][b] = [a][hb]$, that is, $[ab] = [ahb]$. This gives us that $Hab = Hahb$, and so $Ha = Hah$; this implies that $H = Haha^{-1}$, whence $aha^{-1} \in H$. That is, for all $a \in G$ and all $h \in H$, aha^{-1} must be in H; in other words, H must be normal in G. *So we see that in order for the product defined in W to be well-defined, H must be a normal subgroup of G.*

We view this matter of the quotient group in a slightly different way. If A, B are subsets of G, let $AB = \{ab | a \in A, b \in B\}$. If H is a subgroup of G, then $HH \subset H$ is another way of saying that H is closed under the product of G.

Let $G/N = \{Na | a \in G\}$ be the set of all left cosets of the normal subgroup N in G. Using the product of subsets of G as defined above, what is $(Na)(Nb)$? By definition, $(Na)(Nb)$ consists of all elements of the form $(na)(mb)$, where $n, m \in N$, and so

$$(na)(mb) = (nama^{-1})(ab) = n_1 ab,$$

where $n_1 = nama^{-1}$ is in N, since N is normal. Thus $(Na)(Nb) \subset Nab$. On the other hand, if $n \in N$, then

$$n(ab) = (na)(eb) \in (Na)(Nb),$$

so that $Nab \subset (Na)(Nb)$. In short, we have shown that the product —as subsets of G—of Na and Nb is given us by the formula $(Na)(Nb) = Nab$. All the other axioms for G/N, as defined here, to be a group are now readily verified from this product formula.

Another way of seeing that $(Na)(Nb) = Nab$ is to note that by the normality of N, $aN = Na$, hence $(Na)(Nb) = N(aN)b = N(Na)b = NNab = Nab$, since $NN = N$ (because N is a subgroup of G).

However we view G/N—as equivalence classes or as a set of certain subsets of G—we do get a group whose structure is intimately tied to that of G, via the natural homomorphism ψ of G onto G/N.

We shall see very soon how we combine induction and the structure of G/N to get information about G.

When G is a finite group and $N \triangleleft G$, then the number of left cosets of N in G, $i_G(N)$, is given—as the proof of Lagrange's Theorem showed—by $i_G(N) = |G|/|N|$. But this is the order of G/N, which is the set of all the left cosets of N in G. Thus $|G/N| = |G|/|N|$. We state this more formally as

Theorem 2.6.3. If G is a finite group and $N \triangleleft G$, then $|G/N| = |G|/|N|$.

As an application of what we have been talking about here, we shall prove a special case of a theorem that we shall prove in its full generality later. The proof we give—for the abelian case— is not a particularly good one, but it illustrates quite clearly a general technique, that of pulling back information about G/N to get information about G itself.

> *The theorem we are about to prove is due to the great French mathematician A. L. Cauchy (1789–1857), whose most basic contributions were in complex variable theory.*

Theorem 2.6.4 (Cauchy). If G is a finite abelian group of order $|G|$ and p is a prime that divides $|G|$, then G has an element of order p.

Proof. Before getting involved with the proof, we point out to the reader that the theorem is true for *any* finite group. We shall prove it in the general case later, with a proof that will be much more beautiful than the one we are about to give for the special, abelian case.

We proceed by induction on $|G|$. What does this mean precisely? We shall assume the theorem to be true for all abelian groups of order less than $|G|$ and show that this forces the theorem to be true for G. If $|G| = 1$, there is no such p and the theorem is vacuously true. So we have a starting point for our induction.

Suppose that there is a subgroup $(e) \neq N \neq G$; since $|N| < |G|$, if $p \mid |N|$, by our induction hypothesis there would be an element of order p in N, hence in G, and we would be done. So we may suppose that $p \nmid |N|$. Since G is abelian, every subgroup is normal, so we can form G/N. Because $p \mid |G|$ and $p \nmid |N|$, and because $|G/N| = |G|/|N|$, we have that $p \mid |G/N|$. The group G/N is abelian, since G is (Prove!) and since $N \neq (e)$, $|N| > 1$, so $|G/N| = |G|/|N| < |G|$. Thus, again by induction, there exists an element in G/N of order p. This translates into: There exists an $a \in G$ such that $[a]^p = [e]$, but $[a] \neq [e]$. Since $[e] = [a]^p = [a^p]$, we have (from the equivalence defined mod N) that $a^p \in N$, $a \notin N$. So if $m = |N|$, then by Theorem 2.4.5, since $a^p \in N$, $(a^p)^m = e$. So $(a^m)^p = e$. If we could show that $b = a^m \neq e$, then b would be the required element of order p in G. But if $a^m = e$, then $[a]^m = [e]$, and since $[a]$ has order p, $p \mid m$ (see Problem 31 of Section 4). But, by assumption, $p \nmid m = |N|$. *So we are done if G has a nontrivial subgroup.*

But if G has no nontrivial subgroups, it must be cyclic of prime order. (See Problem 16 of Section 3, which you should be able to handle more easily now.) What is this "prime order"? Because $p \mid |G|$, we must have $|G| = p$. But then any element $a \neq e \in G$ satisfies $a^p = e$ and is of order p. This completes the induction, and so proves the theorem. $\square$

We shall have other applications of this kind of group-theoretic argument in the problems.

The notion of a factor group is a very subtle one, and of the greatest importance in the subject. The formation of a new set from an old one by using as elements of this new set subsets of the old one is strange to the neophyte seeing this kind of construction for the first time. So it is worthwhile looking at this whole matter from a variety of points of view. We consider G/N from another angle now.

What are we doing when we form G/N? Sure, we are looking at equivalences classes defined via N. Let's look at it another way. What we are doing is *identifying* two elements in G if they satisfy the relation $ab^{-1} \in N$. In a sense we are blotting out N. So although G/N is *not* a subgroup of G, we can look at it as G, with N blotted out, and two elements as equal if they are equal "up to N."

For instance, in forming Z/N, where Z is the group of integers and N is the set of all multiples of 5 in Z, what we are doing is *identifying* 1 with $6, 11, 16, -4, -9$, and so on, and we are identifying all multiples of 5 with 0. The nice thing about all this is that this identification jibes with addition in Z when we go over to Z/N.

Let's look at a few examples from this point of view.

1. Let $G = \{T_{a,b} \mid a \neq 0, b \text{ real}\}$ (Example 6 of Section 1). Let $N = \{T_{1,b} \mid b \text{ real}\} \subset G$; we saw that $N \lhd G$, so it makes sense to talk about G/N. Now $T_{a,b}$ and $T_{a,0}$ are in the same left coset of N in G, so in G/N we are getting an element by identifying $T_{a,b}$ with $T_{a,0}$. The latter element just depends on a. Moreover, the $T_{a,b}$ multiply according as the first subscript a for $T_{a,b}T_{c,d} = T_{ac,ad+b}$ and if we identify $T_{a,b}$ with $T_{a,0}$, $T_{c,d}$ with $T_{c,0}$, then their product, which is $T_{ac,ad+b}$, is identified with ac. So in G/N multiplication is like that of the group of nonzero real numbers under multiplication, and in some sense (which will be made more precise in the next section) G/N can be identified with this group of real numbers.

2. Let G be the group of real numbers under $+$ and let $\mathbb{Z}$ be the group of integers under $+$. Since G is abelian, $\mathbb{Z} \triangleleft G$, and so we can talk about $G/\mathbb{Z}$. What does $G/\mathbb{Z}$ really look like? *In forming $G/\mathbb{Z}$, we are identifying any two real numbers that differ by an integer.* So 0 is identified with $-1, -2, -3, \ldots$ and $1, 2, 3, \ldots$; $\frac{3}{2}$ is identified with $\frac{1}{2}, \frac{5}{2}, -\frac{1}{2}, -\frac{3}{2}, \ldots$. Every real number a thus has a mate, $\tilde{a}$, where $0 \leq \tilde{a} < 1$. So, in $G/\mathbb{Z}$, the whole real line has been compressed into the unit interval $[0, 1]$. But a little more is true, for we have also identified the end points of this unit interval. So we are bending the unit interval around so that its two end points touch and become one. What do we get this way? A circle, of course! So $G/\mathbb{Z}$ is like a circle, in a sense that can be made precise, and this circle is a group with an appropriate product.

3. Let G be the group of nonzero complex numbers and let $N = \{a \in G \,|\, |a| = 1\}$ which is the circle of radius 1 and center 0 in the complex plane. Then N is a subgroup of G and is normal since G is abelian. In going to G/N we are declaring that any complex number of absolute value 1 will be identified with the real number 1. Now any $a \in G$, in its polar form, can be written as $a = r(\cos\theta + i\sin\theta)$, where $r = |a|$, and $|\cos\theta + i\sin\theta| = 1$. In identifying $\cos\theta + i\sin\theta$ with 1, we are identifying a with r. So in passing to G/N every element is being identified with a positive real number, and this identification jibes with the products in G and in the group of positive real numbers, since $|ab| = |a|\,|b|$. So G/N is in a very real sense (no pun intended) the group of positive real numbers under multiplication.

Problems

1. If G is the group of all nonzero real numbers under multiplication and N is the subgroup of all positive real numbers, write out G/N by exhibiting the cosets of N in G, and construct the multiplication in G / N.

2. If G is the group of nonzero real numbers under multiplication and $N = \{1, -1\}$, show how you can "identify" G/N as the group of all positive real numbers under multiplication. What are the cosets of N in G?

3. If G is the group and $N \triangleleft G$, show that if $\overline{M}$ is a subgroup of G/N and $M = \{a \in G | Na \in \overline{M}\}$, then M is a subgroup of G, and $M \supset N$.

4. If $\overline{M}$ in Problem 3 is normal in G/N, show that the M defined is normal in G.

5. In Problem 3, show that M/N must equal $\overline{M}$.

6. Arguing as in the Example 2, where we identified $G/\mathbb{Z}$ as a circle, where G is the group of reals under $+$ and $\mathbb{Z}$ the integers, consider the following: let $G = \{(a, b) | a, b \text{ real}\}$, where $+$ in G is defined by $(a, b) + (c, d) = (a + c, b + d)$ (so G is the plane), and let $N = \{(a, b) \in G | a, b \text{ are integers}\}$. Show that G/N can be identified as a torus (donut), and so we can define a product on the donut so that it becomes a group. Here, you may think of a torus as the Cartesian product of two circles.

7. If G is a cyclic group and N is a subgroup of G, show that G/N is a cyclic group.

8. If G is an abelian group and N is a subgroup of G, show that G/N is an abelian group.

9. Do Problems 7 and 8 by observing that G/N is a homomorphic image of G.

10. Let G be an abelian group of order $p_1^{a_1} p_2^{a_2} \cdots p_k^{a_k}$, where $p_1, p_2, \ldots, p_k$ are distinct prime numbers. Show that G has subgroups $S_1, S_2, \ldots, S_k$ of orders $p_1^{a_1}, p_2^{a_2}, \ldots, p_k^{a_k}$, respectively. (**Hint:** Use Cauchy's Theorem and pass to a factor group.) This result, which actually holds for all finite groups, is a famous result in group theory known as *Sylow's Theorem*. We prove it in Section 11.

11. If G is a group and $Z(G)$ the center of G, show that if $G/Z(G)$ is cyclic, then G is abelian.

12. If G is a group and $N \triangleleft G$ is such that G/N is abelian, prove that $aba^{-1}b^{-1} \in N$ for all $a, b \in G$.

13. If G is a group and $N \triangleleft G$ is such that

$$aba^{-1}b^{-1} \in N$$

for all $a, b \in G$, prove that G/N is abelian.

14. If G is an abelian group of order $p_1 p_2 \cdots p_k$, where $p_1, p_2, \ldots, p_k$ are distinct primes, prove that G is cyclic. (See Problem 15.)

15. If G is an abelian group and if G has an element of order m and one of order n, where m and n are relatively prime, prove that G has an element of order mn.

16. Let G be an abelian group of order $p^n m$, where p is a prime and $p \nmid m$. Let $P = \{ a \in G | a^{p^k} = e$ for some k depending on $a \}$. Prove that:
   (a) P is a subgroup of G.
   (b) G/P has no elements of order p.
   (c) $|P| = p^n$.

17. Let G be an abelian group of order mn, where m and n are relatively prime. Let $M = \{ a \in G | a^m = e \}$. Prove that:
   (a) M is a subgroup of G.
   (b) G/M has no element, x, other than the identity element, such that $x^m =$ unit element of G/N.

18. Let G be an abelian group (possibly infinite) and let the set $T = \{ a \in G | a^m = e, \ m > 1$ depending on $a \}$. Prove that:
   (a) T is a subgroup of G.
   (b) G/T has no element—other than its identity element—of finite order.

7.   THE HOMOMORPHISM THEOREMS

Let G be a group and φ a homomorphism of G *onto* G'. If K is the kernel of φ, then K is a normal subgroup of G, hence we can form G/K. It is fairly natural to expect that there should be a very close relationship between G' and G/K. The *First Homomorphism Theorem*, which we are about to prove, spells out this relationship in exact detail.

But first let's look back at some of the examples of factor groups in Section 6 to see explicitly what the relationship mentioned above might be.

1. Let $G = \{ T_{a,b} | a \neq 0, \ b$ real$\}$ and let G' be the group of nonzero reals under multiplication. From the product rule of these T's,

namely $T_{a,b}T_{c,d} = T_{ac,ad+b}$, we determined that the mapping $\varphi : G \to G'$ defined by $\varphi(T_{a,b}) = a$ is a homomorphism of G onto G' with kernel $k = \{T_{1,b} | b \text{ real}\}$. On the other hand, in Example 1 of Section 6 we saw that $G/K = \{KT_{a,0} | a \neq 0 \text{ real}\}$. Since

$$(KT_{a,0})(KT_{x,0}) = KT_{ax,0}$$

the mapping of G/K onto G', which sends each $KT_{a,0}$ onto a, is readily seen to be an *isomorphism* of G/K onto G'. Therefore, $G/K \simeq G'$.

2. In Example 3, G was the group of nonzero complex numbers under multiplication and G' the group of all positive real numbers under multiplication. Let $\varphi : G \to G'$ defined by $\varphi(a) = |a|$ for $a \in G$. Then, since $|ab| = |a| |b|$, φ is a homomorphism of G onto G' (can you see why it is onto?). Thus the kernel K of φ is precisely $K = \{a \in G | |a| = 1\}$. But we have already seen that if $|a| = 1$, then a is of the form $\cos\theta + i\sin\theta$. So the set $K = \{\cos\theta + i\sin\theta | 0 \leq \theta < 2\pi\}$. If a is any complex number, then $a = r(\cos\theta + i\sin\theta)$, where $r = |a|$, is the polar form of a. Thus $Ka = Kr(\cos\theta + i\sin\theta) = K(\cos\theta + i\sin\theta)r = Kr$, since $K(\cos\theta + i\sin\theta) = K$ because $\cos\theta + i\sin\theta \in K$. So G/K, whose elements are the cosets Ka, from this discussion, has all its elements of the form Kr, where $r > 0$. The mapping of G/K onto G' defined by sending Kr onto r then defines an *isomorphism* of G/K onto G'. So, here, too, $G/K \simeq G'$.

With this little experience behind us we are ready to make the jump the whole way, namely, to

Theorem 2.7.1 (First Homomorphism Theorem). Let φ be a homomorphism of G onto G' with kernel K. Then $G' \simeq G/K$, the isomorphism between these effected by the map

$$\psi : G/K \to G'$$

defined by $\psi(Ka) = \varphi(a)$.

Proof. The best way to show that G/K and G' are isomorphic is to exhibit explicitly an isomorphism of G/K onto G'. The statement of the theorem suggests what such an isomorphism might be.

So define $\psi: G/K \to G'$ by $\psi(Ka) = \varphi(a)$ for $a \in G$. As usual, our first task is to show that ψ is well defined, that is, to show that if $Ka = Kb$, then $\psi(Ka) = \psi(Kb)$. This boils down to showing that if $Ka = Kb$, then $\varphi(a) = \varphi(b)$. But if $Ka = Kb$, then $a = kb$ for some $k \in K$, hence $\varphi(a) = \varphi(kb) = \varphi(k)\varphi(b)$; since $k \in K$, the kernel of φ, then $\varphi(k) = e'$, the identity element of G', so we get $\varphi(a) = \varphi(b)$. This shows that the mapping ψ is well defined.

Because φ is onto G', given $x \in G'$, then $x = \varphi(a)$ for some $a \in G$, thus $x = \varphi(a) = \psi(Ka)$. This shows that ψ maps G/K onto G'.

Is ψ 1-1? Suppose that $\psi(Ka) = \psi(Kb)$; then $\varphi(a) = \psi(Ka) = \psi(Kb) = \varphi(b)$; therefore, $e' = \varphi(a)\varphi(b)^{-1} = \varphi(a)\varphi(b^{-1}) = \varphi(ab^{-1})$. Because ab^{-1} is thus in the kernel of φ—which is K—we have that $ab^{-1} \in K$. This implies that $Ka = Kb$. In this way ψ is seen to be 1-1.

Finally, is ψ a homomorphism of G/K onto G'? We check: $\psi((Ka)(Kb)) = \psi(Kab) = \varphi(ab) = \varphi(a)\varphi(b) = \psi(Ka)\psi(Kb)$, using that φ is a homomorphism and that $(Ka)(Kb) = Kab$. Consequently, ψ is a homomorphism of G/K onto G', and Theorem 2.7.1 is proved. $\square$

Having talked about the *First* Homomorphism Theorem suggests that there are others. The next one we naturally call the *Second Homomorphism Theorem*.

Theorem 2.7.2 (Second Homomorphism Theorem). Let the map $\varphi: G \to G'$ be a homomorphism of G onto G' with kernel K. If H' is a subgroup of G' and if

$$H = \{a \in G | \varphi(a) \in H'\},$$

then H is a subgroup of G, $H \supset K$, and $H/K \simeq H'$. Finally, if $H' \lhd G'$, then $H \lhd G$.

Proof. We first verify that the H above is a subgroup of G. It is not empty, since $e \in H$. If $a, b \in H$, then $\varphi(a)$, $\varphi(b) \in H'$, hence $\varphi(ab) = \varphi(a)\varphi(b) \in H'$, since H' is a subgroup of G'; this puts ab in H, so H is closed. Further, if $a \in H$, then $\varphi(a) \in H'$, hence $\varphi(a^{-1}) = \varphi(a)^{-1}$ is in H', again since H' is a subgroup of G', whence $a^{-1} \in H$. Therefore, H is a subgroup of G.

Because $\varphi(K) = \{e'\} \subset H'$, where e' is the unit element of G', we have that $K \subset H$. Since $K \triangleleft G$ and $K \subset H$, it follows all the more so that $K \triangleleft H$. The mapping φ restricted to H defines a homomorphism of H onto H' with kernel K. By the First Homomorphism Theorem we get $H/K \approx H'$.

Finally, if $H' \triangleleft G'$ and if $a \in G$, then $\varphi(a)^{-1} H' \varphi(a) \subset H'$, so $\varphi(a^{-1}) H' \varphi(a) \subset H'$. This tells us that $\varphi(a^{-1} H a) \subset H'$, as $a^{-1} H a \subset H$. This proves the normality of H in G. $\square$

By this theorem, there is a 1-1 correspondence between the set of all subgroups H of G that contain K and the set of all subgroups H' of G'. Moreover, this correspondance preserves normality in the sense that if H is normal in G, then H' is normal in G'.

Finally, we go on to the *Third Homomorphism Theorem*, which tells us a little more about the relationship between N and N' when $N' \triangleleft G'$.

Theorem 2.7.3 (Third Homomorphism Theorem). If the map $\varphi: G \to G'$ is a homomorphism of G onto G' with kernel K then, if $N' \triangleleft G'$ and $N = \{a \in G | \varphi(a) \in N'\}$, we conclude that $G/N \simeq G'/N'$. Equivalently, $G/N \simeq (G/K)/(N/K)$.

Proof. Define the mapping $\psi: G \to G'/N'$ by $\psi(a) = N'\varphi(a)$ for every $a \in G$. Since φ is onto G' and every element of G'/N' is a coset of the form $N'x'$, and $x' = \varphi(x)$ for some $x \in G$, we see that ψ maps G onto G'/N'.

Furthermore, ψ is a homomorphism of G onto G'/N', for $\psi(ab) = N'\varphi(ab) = N'\varphi(a)\varphi(b) = (N'\varphi(a))(N'\varphi(b)) = \psi(a)\psi(b)$, since $N' \triangleleft G'$. What is the kernel, M, of ψ? If $a \in M$, then $\psi(a)$ is the unit element of G'/N', that is, $\psi(a) = N'$. On the other hand, by the definition of ψ, $\psi(a) = N'\varphi(a)$. Because $N'\varphi(a) = N'$ we must have $\varphi(a) \in N'$; but this puts a in N, by the very definition of N. Thus $M \subset N$. That $N \subset M$ is easy and is left to the reader. Therefore, $M = N$, so ψ is a homomorphism of G onto G'/N' with kernel N, whence, by the First Homomorphism Theorem, $G/N \simeq G'/N'$.

Finally, again by Theorems 2.7.1 and 2.7.2, $G' \simeq G/K$, $N' \simeq N/K$, which leads us to $G/N \simeq G'/N' \simeq (G/K)/(N/K)$. $\square$

This last equality is highly suggestive; we are sort of "canceling out" the K in the numerator and denominator.

There are other homomorphism theorems, of course. One of the classic and important ones occurs in Problem 5.

Problems

1. Show that $M \supset N$ in the proof of Theorem 2.7.3.

2. Let G be the group of all real-valued functions on the unit interval $[0, 1]$, where we define, for $f, g \in G$, addition by $f + g$ by $(f + g)(x) = f(x) + g(x)$ for every $x \in [0, 1]$. If $N = \{f \in G | f(\frac{1}{4}) = 0\}$, prove that $G/N \simeq$ real numbers under $+$.

3. Let G be the group of nonzero real numbers under multiplication and let $N = \{1, -1\}$. Prove that $G/N \simeq$ positive real numbers under multiplication.

4. If G_1, G_2 are two groups and $G = G_1 \times G_2 = \{(a, b) | a \in G_1, b \in G_2\}$, where we define $(a, b)(c, d) = (ac, bd)$, show that:
   (a) $N = \{(a, e_2) | a \in G_1\}$, where e_2 is the unit element of G_2, is a normal subgroup of G.
   (b) $N \simeq G_1$.
   (c) $G/N \simeq G_2$.

5. Let G be a group, H a subgroup of G, and $N \triangleleft G$. Let the set $HN = \{hn | h \in H, n \in N\}$. Prove that:
   (a) $H \cap N \triangleleft H$.
   (b) HN is a subgroup of G.
   (c) $N \subset HN$ and $N \triangleleft HN$.
   (d) $(HN)/N \simeq H/(H \cap N)$.

6. If G is a group and $N \triangleleft G$, show that if $a \in G$ has finite order $o(a)$, then Na in G/N has finite order m, where $m | o(a)$. (Prove this by using the homomorphism of G onto G/N.)

7. If φ is a homomorphism of G onto G' and $N \triangleleft G$, show that $\varphi(N) \triangleleft G'$.

8. CAUCHY'S THEOREM

In Theorem 2.6.4—Cauchy's Theorem—we proved that if a prime p divides the order of a finite *abelian* group G, then G contains an element of order p. We did point out there that Cauchy's Theorem is true even if the group is not abelian. We shall give a very neat proof of this here; this proof is due to McKay.

We return for a moment to set theory, doing something that we mentioned in the problems in Section 4.

Let S be a set, $f \in A(S)$, and define a relation on S as follows: $s \sim t$ if $t = f^i(s)$ for some integer i (i can be positive, negative, or zero). We leave it to the reader as a problem that this does indeed define an equivalence relation on S. The equivalence class of s, $[s]$, is called the *orbit* of s under f. So S is the disjoint union of the orbits of its elements.

When f is of order p, p a prime, we can say something about the size of the orbits under f; those of the readers who solved Problem 34 of Section 4 already know the result. We prove it here to put it on the record officially.

[If $f^k(s) = s$, of course $f^{tk}(s) = s$ for every integer t. (Prove!)]

Lemma 2.8.1. If $f \in A(S)$ is of order p, p a prime, then the orbit of any element of S under f has 1 or p elements.

Proof. Let $s \in S$; if $f(s) = s$, then the orbit of s under f consists merely of s itself, so has one element. Suppose then that $f(s) \neq s$. Consider the elements $s, f(s), f^2(s), \ldots, f^{p-1}(s)$; we claim that these p elements are distinct and constitute the orbit of s under f. If not, then $f^i(s) = f^j(s)$ for some $0 \leq i < j \leq p - 1$, which gives us that $f^{j-i}(s) = s$. Let $m = j - i$; then $0 < m \leq p - 1$ and $f^m(s) = s$. But $f^p(s) = s$ and since $p \nmid m$, $ap + bm = 1$ for some integers a and b. Thus $f^1(s) = f^{ap+bm}(s) = f^{ap}(f^{bm}(s)) = f^{ap}(s) = s$, since $f^m(s) = f^p(s) = s$. This contradicts that $f(s) \neq s$. Thus the orbit of s under f consists of s, $f(s), f^2(s), \ldots, f^{p-1}(s)$, so has p elements. $\square$

We now give McKay's proof of Cauchy's Theorem.

Theorem 2.8.2 (Cauchy).   If p is a prime and p divides the order of G, then G contains an element of order p.

Proof.   If $p = 2$, the result amounts to Problem 18 in Section 1. Assume that $p \neq 2$. Let S be the set of all *ordered* p-tuples $(a_1, a_2, \ldots, a_{p-1}, a_p)$, where $a_1, \ldots, a_p$ are in G and where $a_1 a_2 \ldots, a_{p-1} a_p = e$. We claim that S has n^{p-1} elements where $n = |G|$. Why? We can choose $a_1 \ldots, a_{p-1}$ arbitrarily in G, and by putting $a_p = (a_1 a_2 \ldots, a_{p-1})^{-1}$ the p-tuple $(a_1 a_2 \ldots, a_{p-1}, a_p)$, then satisfies

$$a_1 a_2 \cdots a_{p-1} a_p = a_1 a_2 \cdots a_{p-1}(a_1 a_2 \cdots a_{p-1})^{-1} = e,$$

so is in S. Thus S has n^{p-1} elements.

Note that if $a_1 a_2 \cdots a_{p-1} a_p = e$, then $a_p a_1 a_2 \cdots a_{p-1} = e$ (for if $xy = e$ in a group, then $yx = e$). So the mapping $f: S \to S$ defined by $f(a_1, \ldots, a_p) = (a_p, a_1, a_2, \ldots, a_{p-1})$ is in $A(S)$. Note that $f \neq i$, the identity map on S, and that $f^p = i$, so f is of order p.

If the orbit of s under f has one element, then $f(s) = s$. On the other hand, if $f(s) \neq s$, we claim that the orbit of s under f consists precisely of p distinct elements; this we have by Lemma 2.8.1. Now when is $f(s) \neq s$? We claim that $f(s) \neq s$ if and only if when $s = (a_1, a_2, \ldots, a_p)$, then for some $i \neq j$, $a_i \neq a_j$. (We leave this to the reader.) So $f(s) = s$ if and only if $s = (a, a, \ldots, a)$ for some $a \in G$.

Let m be the number of $s \in S$ such that $f(s) = s$; since for $s = (e, e, \ldots, e)$, $f(s) = s$, we know that $m \geq 1$. On the other hand, if $f(s) \neq s$, the orbit of s consists of p elements, and these orbits are disjoint, for they are equivalence classes. If there are k such orbits where $f(s) \neq s$, we get that $n^{p-1} = m + kp$, for we have accounted this way for every element of S.

But $p | n$ by assumption and $p | (kp)$, so we must have $p | m$, since $m = n^{p-1} - kp$. Because $m \neq 0$ and $p | m$, we get that $m > 1$. But this says that there is an $s = (a, a, \ldots, a) \neq (e, e, \ldots, e)$ in S; from the definition of S this implies that $a^p = e$. Since $a \neq e$, a is the required element of order p. $\square$

Note that the proof tells us that the number of solutions in G of $x^p = e$ is a positive multiple of p.

We strongly urge the reader who feels uncomfortable with the proof just given to carry out its details for $p = 3$. In this case the action of f on S becomes clear and our assertions about this action can be checked explicitly.

Cauchy's Theorem has many consequences. We shall present one of these, in which we determine completely the nature of certain groups of order pq, where p and q are distinct primes, soon. Other consequences will be found in the problem set to follow, and in later material on groups.

> **Lemma 2.8.3.** Let G be a group of order pq, where p, q are primes and $p > q$. If $a \in G$ is of order p and A is the subgroup of G generated by a, then $A \triangleleft G$.

Proof. We claim that A is the *only* subgroup of G of order p. For suppose that B is another subgroup of order p. Consider the set $AB = \{xy | x \in A, y \in B\}$; we claim that AB has p^2 distinct elements. For suppose that $xy = uv$ where $x, u \in A$, $y, v \in B$; then $u^{-1}x = vy^{-1}$. But $u^{-1}x \in A$, $vy^{-1} \in B$, and since $u^{-1}x = vy^{-1}$, we have $u^{-1}x \in A \cap B$. Since $B \neq A$ and $A \cap B$ is a subgroup of A and A is of prime order, we are forced to conclude that $A \cap B = (e)$ and so $u^{-1}x = e$, that is, $u = x$. Similarly, $v = y$. Thus the number of distinct elements in AB is p^2. But all these elements are in G, which has only $pq < p^2$ elements (since $p > q$). With this contradiction we see that $B = A$ and A is the only subgroup of order p in G. But if $x \in G$, $B = x^{-1}Ax$ is a subgroup of G of order p, in consequence of which we conclude that $x^{-1}Ax = A$; hence $A \triangleleft G$. $\square$

> **Corollary.** If G, a are as in Lemma 2.8.3 and if $x \in G$, then $x^{-1}ax = a^i$, where $0 < i < p$, for some i (depending on x).

Proof. Since $e \neq a \in A$ and $x^{-1}Ax = A$, $x^{-1}ax \in A$. But every element of A is of the form a^i, $0 \leq i < p$, and $x^{-1}ax \neq e$. In consequence, $x^{-1}ax = a^i$, where $0 < i < p$. $\square$

We now prove a result of a different flavor.

Lemma 2.8.4.   If $a \in G$ is of order m and $b \in G$ is of order n, where m and n are relatively prime, then, if $ab = ba$, then $c = ab$ is of order mn.

Proof.   Suppose that A is the subgroup generated by a and B that generated by b. Because $|A| = m$ and $|B| = n$ and $(m, n) = 1$, we get $A \cap B = (e)$, which follows from Lagrange's theorem, for $|A \cap B| \, | n$ and $|A \cap B| \, | m$.

Suppose that $c^i = e$, where $i > 0$; thus $(ab)^i = e$. Since $ab = ba$, $e = (ab)^i = a^i b^i$; this tells us that $a^i = b^{-i} \in A \cap B = (e)$. So $a^i = e$, whence $m | i$, and $b^i = e$, whence $n | i$. Because $(m, n) = 1$ and m and n both divide i, mn divides i. So $i \geq mn$. Since $(ab)^{mn} = a^{mn} b^{mn} = e$, we see that mn is the smallest positive integer i such that $(ab)^i = e$. This says that ab is of order mn, as claimed in the lemma. $\square$

Before considering the more general case of groups of order pq, let's look at a special case, namely, a group G of order 15. By Cauchy's Theorem, G has elements b of order 3 and a of order 5. By the Corollary to Lemma 2.83, $b^{-1}ab = a^i$, where $0 < i < 5$. Thus

$$b^{-2}ab^2 = b^{-1}(b^{-1}ab)b = b^{-1}a^i b = (b^{-1}ab)^i = (a^i)^i = a^{i^2},$$

and similarly, $b^{-3}ab^3 = a^{i^3}$. But $b^3 = e$, so we get $a^{i^3} = a$, whence $a^{i^3-1} = e$. Since a is of order 5, 5 must divide $i^3 - 1$, that is, $i^3 \equiv 1(5)$. However, by Fermat's Theorem (Corollary to Theorem 2.4.8), $i^4 \equiv 1(5)$. These two equations for i tell us that $i \equiv 1(5)$, so, since $0 < i < 5$, $i = 1$. In short, $b^{-1}ab = a^i = a$, which means that $ab = ba$. Since a is of order 5 and b of order 3, by Lemma 2.8.3, $c = ab$ is of order 15. This means that the 15 powers $e = c^0, c, c^2, \ldots, c^{14}$ are distinct, so must sweep out all of G. In a word, G *must be cyclic.*

The argument given for 15 could have been made shorter, but the form in which we did it is the exact prototype for the proof of the more general

Theorem 2.8.5.   Let G be a group of order pq, where p, q are primes and $p > q$. Then, if $q \nmid p - 1$, G must be cyclic.

Proof. By Cauchy's Theorem, G has an element a of order p and an element b of order q. By the Corollary to Lemma 2.8.2, $b^{-1}ab = a^i$ for some i with $0 < i < p$. Thus $b^{-r}ab^r = a^{i^r}$ for all $r \geq 0$ (Prove!), and so $b^{-q}ab^q = a^{i^q}$. But $b^q = e$; therefore, $a^{i^q} = a$ and so $a^{i^q - 1} = e$. Because a is of order p, we conclude that $p | i^q - 1$, which is to say, $i^q \equiv 1(p)$. However, by Fermat's Theorem, $i^{p-1} \equiv 1(p)$; since $q \nmid p - 1$, we conclude that $i \equiv 1(p)$, and since $0 < i < p$, $i = 1$ follows. Therefore, $b^{-1}ab = a^i = a$, hence $ab = ba$. By Lemma 2.8.3, $c = ab$ has order pq, so the powers of c sweep out all of G. Thus G is cyclic, and the theorem is proved. $\square$

Problems

Middle-Level Problems

1. In the proof of Theorem 2.8.2, show that if some two entries in $s = (a_1, a_2, \ldots, a_p)$ are different, then $f(s) \neq s$, and the orbit of s under f has p elements.

2. Prove that a group of order 35 is cyclic.

3. Using the result of Problem 40 of Section 5, give another proof of Lemma 2.8.3. (**Hint:** Use for H a subgroup of order p.)

4. Construct a nonabelian group of order 21. (**Hint:** Assume that $a^3 = e$, $b^7 = e$ and find some i such that $a^{-1}ba = a^i \neq a$, which is consistent with the relations $a^3 = b^7 = e$.)

5. Let G be a group of order $p^n m$, where p is prime and $p \nmid m$. Suppose that G has a normal subgroup of P of order p^n. Prove that $\theta(P) = P$ for every automorphism θ of G.

6. Let G be a finite group and suppose A, B subgroups of G such that $|A| > \sqrt{|G|}$ and $|B| > \sqrt{|G|}$. Prove that $A \cap B \neq (e)$.

7. If G is a a group and A, B subgroups of orders m, n, respectively, where m and n are relatively prime, prove that the subset of G $AB = \{ab | a \in A, b \in B\}$ has mn distinct elements.

8. Prove that a group of order 99 has a nontrivial normal subgroup.

9. Prove that a group of order 42 has a nontrivial normal subgroup.

10. From the result of Problem 9, prove that a group of order 42 has a normal subgroup of order 21.

Harder Problems

11. If G is a group and A, B finite subgroups of G, prove that the set $AB = \{ab | a \in A, b \in B\}$ has $(|A| |B|)/|A \cap B|$ distinct elements.

12. Prove that any two nonabelian groups of order 21 are isomorphic. (See Problem 4.)

Very Hard Problems

13. Using the fact that any group of order 9 is abelian, prove that any group of order 99 is abelian.

14. Let $p > q$ be two primes such that $q|p - 1$. Prove that there exists a nonabelian group of order pq. (**Hint:** Use the result of Problem 40 of Section 4, namely that U_p is cyclic if p is a prime, and the idea needed to do Problem 4 above.)

15. Prove that if $p > q$ are two primes such that $q|p - 1$, then any two nonabelian groups of order pq are isomorphic.

9.   DIRECT PRODUCTS

In several of the problems and examples that appeared earlier, we went through the following construction: If G_1, G_2 are two groups, then $G = G_1 \times G_2$ is the set of all ordered pairs (a, b), where $a \in G_1$ and $b \in G_2$ and where the product was defined *component-wise* via $(a_1, b_1)(a_2, b_2) = (a_1 a_2, b_1 b_2)$, the products in each component being carried out in the respective groups G_1 and G_2. We should like to formalize this procedure here.

> **Definition.** If $G_1, G_2, \ldots, G_n$ are n groups, then their (*external*) *direct product* $G_1 \times G_2 \times G_3 \times \cdots \times G_n$ is the set of all ordered n-tuples $(a_1, a_2, \ldots, a_n)$ where $a_i \in G_i$, for $i = 1, 2, \ldots, n$, and where the product in $G_1 \times G_2 \times \cdots \times G_n$ is defined component-wise, that is,
>
> $$(a_1, a_2, \ldots, a_n)(b_1, b_2, \ldots, b_n) = (a_1 b_1, a_2 b_2, \ldots, a_n b_n).$$

That $G = G_1 \times G_2 \times \cdots \times G_n$ is a group is immediate, with $(e_1, e_2, \ldots, e_n)$ as its unit element, where e_i is the unit element of G_i, and where $(a_1, a_2, \ldots, a_n)^{-1} = (a_1^{-1}, a_2^{-1}, \ldots, a_n^{-1})$.

G is merely the Cartesian product of the groups $G_1, G_2, \ldots, G_n$ with a product defined in G by component-wise multiplication. We call it *external*, since the groups $G_1, G_2, \ldots, G_n$ are any groups, with no relation necessarily holding among them.

Consider the subsets $\overline{G}_i \subset G_1 \times G_2 \times \cdots \times G_n = G$, where

$$\overline{G}_i = \{(e_1, \ldots, e_{i-1}, a_i, e_{i+1}, \ldots, e_n)|a_i \in G_i\};$$

in other words, $\overline{G}_i$ consists of all n-tuples where in the ith component any element of G_i can occur and where every other component is the identity element. Clearly, $\overline{G}_i$ is a group and is isomorphic to G_i by the isomorphism $\pi_i \colon \overline{G}_i \to G_i$ defined by $\pi_i(e_1, e_2, \ldots, a_i, \ldots, e_n) = a_i$. Furthermore, not only is $\overline{G}_i$ a subgroup of G but $\overline{G}_i \triangleleft G$. (Prove!)

Given any element $a = (a_1, a_2, \ldots, a_n) \in G$, then

$$a = (a_1, e_2, \ldots, e_n)(e_1, a_2, e_3, \ldots, e_n) \cdots (e_1, e_2, \ldots, e_{n-1}, a_n);$$

that is, every $a \in G$ can be written as $a = \overline{a}_1 \overline{a}_2 \cdots \overline{a}_n$, where each $\overline{a}_i \in \overline{G}_i$. Moreover, a can be written in this way in a unique manner, that is, if $a = \overline{a}_1 \overline{a}_2 \cdots \overline{a}_n = \overline{b}_1 \overline{b}_2 \cdots \overline{b}_n$, where the $\overline{a}_i \in \overline{G}_i$ and $\overline{b}_i \in \overline{G}_i$, then $\overline{a}_1 = \overline{b}_1, \ldots, \overline{a}_n = \overline{b}_n$. So G is built up from certain normal subgroups, the $\overline{G}_i$, as $G = \overline{G}_1 \overline{G}_2 \cdots \overline{G}_n$ in such a way that every element $a \in G$ has a *unique* representation in the form $a = \overline{a}_1 \overline{a}_2 \cdots \overline{a}_n$ with $\overline{a}_i \in \overline{G}_i$.

This motivates the following

Definition. The group G is said to be the (*internal*) *direct product* of its *normal* subgroups $N_1, N_2, \ldots, N_n$ if every $a \in G$ has the *unique* representation in the form $a = a_1 a_2 \cdots a_n$, where each $a_i \in N_i$ for $i = 1, 2, \ldots, n$.

From what we have discussed above we have the

Lemma 2.9.1. If $G = G_1 \times G_2 \times \cdots \times G_n$ is the external direct product of $G_1, G_2, \ldots, G_n$, then G is the internal direct product of the normal subgroups $\overline{G}_1, \overline{G}_2, \ldots, \overline{G}_n$ defined above.

We want to go in the other direction, namely to prove that if G is the internal direct product of its normal subgroups $N_1, N_2, \ldots, N_n$, then G is isomorphic to $N_1 \times N_2 \times \cdots \times N_n$. To do so, we first get some preliminary results.

The result we are about to prove has already occured as Problem 20, Section 5. For the sake of completeness we prove it here.

Lemma 2.9.2. Let G be a group, M, N normal subgroups of G such that $M \cap N = (e)$. Then, given $m \in M$ and $n \in N$, $mn = nm$.

Proof. Consider the element $a = mnm^{-1}n^{-1}$. Viewing a as bracketed one way, $a = (mnm^{-1})n^{-1}$; then, since $N \triangleleft G$ and $n \in N$, $mnm^{-1} \in N$, so $a = (mnm^{-1})n^{-1}$ is also in N. Now bracket a in the other way, $a = m(nm^{-1}n^{-1})$; since $M \triangleleft G$ and $m^{-1} \in M$, we have $nm^{-1}n^{-1} \in M$ and so $a = m(nm^{-1}n^{-1}) \in M$. Thus $a \in M \cap N = (e)$, which is to say, $mnm^{-1}n^{-1} = e$. This gives us that $mn = nm$, as required. $\square$

If G is the internal direct product of the normal subgroups $N_1, N_2, \ldots, N_n$, we claim that $N_i \cap N_j = (e)$ for $i \neq j$. For suppose that $a \in N_i \cap N_j$; then $a = e \cdot e \cdots eae \cdots e$, where the a occurs in the ith place. This gives us one representation of a in $G = N_1 N_2 \cdots N_n$. On the other hand, $a = e \cdot e \cdots e \cdot a \cdot e \cdots e$, where the a occurs in the jth place, so a has the second representation as an element of $N_1 N_2 \cdots N_n$. By the uniqueness of the representation, we get $a = e$, and so $N_i \cap N_j = (e)$.

Perhaps things would be clearer if we do it for $n = 2$. So suppose that $N_1 \triangleleft G$, $N_2 \triangleleft G$, and every element $a \in G$ has a unique representation as $a = a_1 a_2$, where $a_1 \in N_1$, $a_2 \in N_2$. Suppose that $a \in N_1 \cap N_2$; then $a = a \cdot e$ is a representation of $a = a_1 a_2$ with $a_1 = a \in N_1$, $a_2 = e \in N_2$. However $a = ea$, so $a = b_1 b_2$, where $b_1 = e \in N_1$, $b_2 = a \in N_2$. By the uniqueness of the representation we must have $a_1 = b_1$, that is, $a = e$. So $N_1 \cap N_2 = (e)$.

The argument given above for $N_1, \ldots, N_n$ is the same argument as that given for $n = 2$, but perhaps is less transparent. At any rate we have proved

Lemma 2.9.3. If G is the internal direct product of its normal subgroups $N_1, N_2, \ldots, N_n$, then, for $i \neq j$, $N_i \cap N_j = (e)$.

Corollary. If G is as in Lemma 2.9.3, then if $i \neq j$ and $a_i \in N_j$ and $a_j \in N_j$, we have $a_i a_j = a_j a_i$.

Proof. By Lemma 2.9.3, $N_i \cap N_j = (e)$ for $i \neq j$. Since the N's are normal in G, by Lemma 2.9.2 we have that any element in N_i *commutes* with any element in N_j, that is, $a_i a_j = a_j a_i$ for $a_i \in N_i$, $a_j \in N_j$. $\square$

With these preliminaries out of the way we can now prove

Theorem 2.9.4. If G is the internal direct product of its normal subgroups $N_1, N_2, \ldots, N_n$, then G is isomorphic to $N_1 \times N_2 \times \cdots \times N_n$, the external direct product of N_1, $N_2, \ldots, N_n$.

Proof. Define the mapping ψ from $N_1 \times N_2 \times \cdots \times N_n$ to G by $\psi((a_1, a_2, \ldots, a_n)) = a_1 a_2 \cdots a_n$. Since every element, a, in G has a representation $a = a_1 a_2 \cdots a_n$, with the $a_i \in N_i$, we have that the mapping ψ is onto. We assert that it is also 1-1. For if $\psi((a_1, a_2, \ldots, a_n)) = \psi((b_1, b_2, \ldots, b_n))$, then by the definition of ψ, $a_1 a_2 \cdots a_n = b_1 b_2 \cdots b_n$. By the uniqueness of the representation of an element in this form we deduce that $a_1 = b_1$, $a_2 = b_2, \ldots, a_n = b_n$. Hence ψ is 1-1.

All that remains is to show that ψ is a homomorphism. So, consider

$$\psi((a_1, a_2, \ldots, a_n)(b_1, b_2, \ldots, b_n)) = \psi((a_1 b_1, a_2 b_2, \ldots, a_n b_n))$$

$$= (a_1 b_1)(a_2 b_2) \cdots (a_n b_n)$$

$$= a_1 b_1 a_2 b_2 \cdots a_n b_n.$$

Since $b_1 \in N_1$, it commutes with a_i, b_i for $i > 1$ by the Corollary to Lemma 2.9.3. So we can pull the b_1 across all the elements to the right of it to get $a_1 b_1 a_2 b_2 \cdots a_n b_n = a_1 a_2 b_2 a_3 b_3 \cdots a_n b_n b_1$. Now repeat this procedure with b_2, and so on, to get that $a_1 b_1 a_2 b_2 \cdots a_n b_n = (a_1 a_2 \cdots a_n)(b_1 b_2 \cdots b_n)$. Thus

$$\psi((a_1, a_2, \ldots, a_n)(b_1, b_2, \ldots, b_n)) = a_1 b_1 a_2 b_2 \cdots a_n b_n$$

$$= (a_1 a_2 \cdots a_n)(b_1 b_2 \cdots b_n)$$

$$= \psi((a_1, a_2, \ldots, a_n))\psi((b_1 b_2, \ldots, b_n)).$$

In other words, ψ is a homomorphism.

With this the proof of Theorem 2.9.4 is complete. $\square$

In view of the result of Theorem 2.9.4 we drop the adjectives "internal" and "external" and merely speak about the "direct product."

The objective is often to show that a given group is the direct product of certain normal subgroups. If one can do this, the structure of the group can be completely determined if we happen to know those of the normal subgroups.

Problems

1. If G_1 and G_2 are groups, prove that $G_1 \times G_2 \simeq G_2 \times G_1$.

2. If G_1 and G_2 are cyclic groups of orders m and n, respectively, prove that $G_1 \times G_2$ is cyclic if and only if m and n are relatively prime.

3. Let G be a group, $A = G \times G$. In A let $T = \{(g, g) | g \in G\}$.
  (a) Prove that $T \simeq G$.
  (b) Prove that $T \triangleleft A$ if and only if G is abelian.

4. Let G be an abelian group of order $p_1^{m_1} p_2^{m_2} \cdots p_k^{m_k}$, where $p_1, p_2, \ldots, p_k$ are distinct primes and $m_1 > 0, m_2 > 0, \ldots, m_k > 0$. By Problem 10 of Section 6, for each i, G has a subgroup P_i of order $p_i^{m_i}$. Show that $G \simeq P_1 \times P_2 \times \cdots \times P_k$.

5. Let G be a finite group, $N_1, N_2, \ldots, N_k$ normal subgroups of G such that $G = N_1 N_2 \cdots N_k$ and $|G| = |N_1| |N_2| \cdots |N_k|$. Prove that G is the direct product of $N_1, N_2, \ldots, N_k$.

6. Let G be a group, $N_1, N_2, \ldots, N_k$ normal subgroups of G such that:
   1. $G = N_1 N_2 \cdots N_k$.
   2. For each i, $N_i \cap (N_1 N_2 \cdots N_{i-1} N_{i+1} \cdots N_k) = (e)$.
   Prove that G is the direct product of $N_1, N_2, \ldots, N_k$.

10. FINITE ABELIAN GROUPS (OPTIONAL)

We have just finished discussing the idea of the direct product of groups. If we were to leave that topic at the point where we ended, it might seem like a nice little construction, but so what? To give some more substance to it, we should prove at least one theorem which says that a group satisfying a certain condition is the direct product of some particularly easy groups. Fortunately, such a class of groups exists, the finite abelian groups. What we shall prove is that any finite abelian group is the direct product of cyclic groups. This reduces most questions about finite abelian groups to questions about cyclic groups, a reduction that often allows us to get complete answers to these questions.

The situation of the structure of finite abelian groups is really a special case of some wider and deeper theorems. To consider these would be going too far afield, especially since the story for finite abelian groups is so important in its own right. The theorem we shall prove is called the *Fundamental Theorem on Finite Abelian Groups*, and rightfully so.

Before getting down to the actual details of the proof, we should like to give a quick sketch of how we shall go about proving the theorem.

Our first step will be to reduce the problem from any finite abelian group to one whose order is p^n, where p is a prime. This step will be fairly easy to carry out, and since the group will have order involving just one prime, the details of the proof will not be cluttered with elements whose orders are somewhat complicated.

So we shall focus on groups of order p^n. Let G be an abelian group of order p^n. We want to show that there exist cyclic subgroups of G, $A_1, A_2, \ldots, A_k$, such that every element $x \in G$ can be written as $x = b_1 b_2 \cdots b_k$, where each $b_i \in A_i$, in a unique way. Otherwise put, since each A_i is cyclic and generated by a_i, say, we want to show that $x = a_1^{m_1} a_2^{m_2} \cdots a_k^{m_k}$, where the elements $a_i^{m_i}$ are unique.

A difficulty appears right away, for there is not just one choice for these elements $a_1, \ldots, a_k$. For instance, if G is the abelian group of order 4 with elements e, a, b, ab, where $a^2 = b^2 = e$ and $ab = ba$, then we can see that if A, B, C are the cyclic subgroups generated by a, b, and ab, respectively, then $G = A \times B = A \times C = B \times C$. So there is a lack of uniqueness in the choice of the a_i. How to get around this?

What we need is a mechanism for picking a_1 and which, when applied after we have picked a_1, will allow us to pick a_2, and so on. What should this mechanism be? Our control on the elements of G lies only in specifying their orders. It is the order of the element—when properly used—that will give us the means to prove the theorem.

Suppose that $G = A_1 \times A_2 \times \cdots A_k$, where $|G| = p^n$ and the A's have been numbered, so that $|A_i| = p^{n_i}$ and $n_1 \geq n_2 \geq \cdots \geq n_k$, and each A_i is cyclic generated by a_i. If this were so and $x = a_1^{m_1} \cdots a_k^{m_k}$, then

$$x^{p^{n_1}} = \left(a_1^{m_1} \cdots a_k^{m_k} \right)^{p^{n_1}} = a_1^{m_1 p^{n_1}} a_2^{m_2 p^{n_1}} \cdots a_k^{m_k p^{n_1}};$$

because $n_1 \geq n_i$, $p^{n_i} | p^{n_1}$, so since every $a_i^{m_i p^{n_i}} = e$, thus $x^{p^{n_1}} = e$. *In other words, a_1 should be an element of G whose order is as large as it can possibly be.* Fine, we can now pick a_1. What do we do for a_2? If $\bar{G} = G/A_1$, then to get the first element needed to represent $\bar{G}$ as a direct product of cyclic groups, we should pick an element in $\bar{G}$ whose order is maximal. What does this translate into in G itself? We want an element a_2 such that a_2 *requires as high a power as possible to fall into* A_1. So that will be the road to the selection of the second element; however, if we pick an element b_2 with this property, it may not do the trick; we may have to adapt it so that it will. The doing of all this is the technical part of the argument and does go through. Then one repeats it appropriately to find an element a_3, and so on.

This is the procedure we shall be going through to prove the theorem. But to smooth out these successive choices of $a_1, a_2, \ldots$, we shall use an induction argument and some subsidiary preliminary results.

With this sketch as guide we hope the proof of the theorem will make sense to the reader. One should not confuse the basic idea in the proof

—which is quite reasonable—with the technical details, which may cloud the issue. So we now begin to fill in the details of the sketch of the proof that we outlined above.

Lemma 2.10.1. Let G be a finite abelian group of order mn, where m and n are relatively prime. If $M = \{x \in G | x^m = e\}$ and $N = \{x \in G | x^n = e\}$, then $G = M \times N$. Moreover, if neither m nor n is 1, then $M \neq (e)$ and $N \neq (e)$.

Proof. The sets M and N defined in the assertion above are quickly seen to be subgroups of G. Moreover, if $m \neq 1$, then by Cauchy's Theorem (Theorem 2.6.4) we readily obtain $M \neq (e)$, and similarly if $n \neq 1$, that $N \neq (e)$. Furthermore, since $M \cap N$ is a subgroup of both M and N, by Lagrange's Theorem, $|M \cap N|$ divides $|M| = m$ and $|N| = n$. Because m and n are relatively prime, we obtain $|M \cap N| = 1$, hence $M \cap N = (e)$.

To finish the proof, we need to show that $G = MN$ and $G = M \times N$. Since m and n are relatively prime, there exist integers r and s such that $rm + sn = 1$. If $a \in G$, then $a = a^1 = a^{sn+rm} = a^{sn}a^{rm}$; since $(a^{sn})^m = a^{snm} = e$ we have that $a^{sn} \in M$. Similarly, $a^{rm} \in N$. Thus $a = a^{sn}a^{rm}$ is in MN. In this way $G = MN$. We leave the final touch, namely, that $G = M \times N$, to the reader. $\square$

An immediate corollary is the

Corollary. Let G be a finite abelian group and let p be a prime such that $p | |G|$. Then $G = P \times T$ for some subgroups P and T, where $|P| = p^m$, $m > 0$, and $|T|$ is not divisible by p.

Proof. Let $P = \{x \in G | x^{p^s} = e$ for some $s\}$ and let the subset $T = \{x \in G | x^t = e$ for t relatively prime to $p\}$. By Lemma 2.10.1, $G = P \times T$ and $P \neq (e)$. Since every element in P has order a power of p, invoking Cauchy's Theorem we obtain that $|P| = p^m$. $\square$

It is easy to see that $p \nmid |T|$ by making use of Lagrange's Theorem. Thus we really have that P is not merely some subgroup of G but is what is called a p-Sylow subgroup of G. (See Section 11).

We now come to the key step in the proof of the theorem we seek. The proof is a little difficult, but once we have this result the rest will be easy.

Theorem 2.10.2. Let G be an abelian group of order p^n, p a prime, and let $a \in G$ have maximal order of all the elements in G. Then $G = A \times Q$, where A is the cyclic subgroup generated by a.

Proof. We proceed by induction on n. If $n = 1$, then $|G| = p$ and G is already a cyclic group generated by any $a \neq e$ in G.

We suppose the theorem to be true for all $m < n$. If there exists an element $b \in G$ such that $b \notin A = (a)$ and $b^p = e$, we show that the theorem is correct. Let $B = (b)$, the subgroup of G generated by b; thus $A \cap B = (e)$.

Let $\bar{G} = G/B$; by assumption $B \neq (e)$, hence $|\bar{G}| < |G|$. In $\bar{G}$, what is the order of $\bar{a} = Ba$? We claim that $o(\bar{a}) = o(a)$. To begin with, we know that $o(\bar{a})|o(a)$ for $a^{o(a)} = e$, hence $\bar{a}^{o(a)} = (Ba)^{o(a)} = Ba^{o(a)} = B$. On the other hand, $\bar{a}^{o(\bar{a})} = \bar{e}$, so $a^{o(\bar{a})} \in B$, and since $a^{o(\bar{a})} \in A$, we see that $a^{o(\bar{a})} \in A \cap B = (e)$, whence $a^{o(\bar{a})} = e$. This tells us that $o(a)|o(\bar{a})$. Hence $o(a) = o(\bar{a})$.

Since $\bar{a}$ is an element of maximal order in $\bar{G}$, by the induction we know that $\bar{G} = (\bar{a}) \times T$ for some subgroup T of $\bar{G}$. By the Second Homomorphism Theorem we also know that $T = Q/B$ for some subgroup Q of G. We claim that $G = A \times Q$. If not, then $A \cap Q \neq (e)$; let $u \in A \cap Q$. Then $u = a^i \in Q$, so $\bar{a}^i \in Q/B = T$, and since $(\bar{a}) \cap T = (\bar{e})$, we have that $\bar{a}^i = \bar{e}$, hence $a^i \in B$. But $a^i \in A$ and $A \cap B = (e)$; therefore, $a^i = e$, which is to say, $u = a^i = e$. Therefore, $A \cap Q = (e)$ and we obtain that $G = A \times Q$.

Suppose, then, that there is *no* element b in G, b not in A, such that $b^p = e$. We claim that this forces $G = A = (a)$, in which case G is a cyclic group. Suppose that $G \neq A$; let $x \in G$, $x \notin A$ have smallest possible order. Because $o(x^p) < o(x)$, we have, by our choice of x, that $x^p \in A$, hence $x^p = a^i$. We claim that $p|i$. Suppose not. Let $m = o(a)$

$= p^s$; because a is of maximal order in G, then, for any $c \in G$, $o(c) = p^r \le o(a) = p^s$, hence $r \le s$, and so $o(c)|o(a) = m$. Therefore, $c^m = e$.

Now $x^p = a^i$ and we supposed that $p \nmid i$; hence $m \nmid im/p$, and so $a^{im/p} \ne e$. But $x^m = (x^p)^{m/p} = (a^i)^{m/p} = a^{im/p} \ne e$, contradicting the fact observed above that $x^m = e$. Thus $p|i$, so $i = jp$ and $x^p = a^i = a^{jp}$. Let $y = a^{-j}x$; then $y \notin A$, since $x \notin A$ and $a^{-j} \in A$; moreover, $y^p = (a^{-j}x)^p = a^{-jp}x^p = e$. But this puts us back in the situation discussed above, where there exists an $x \in G$, $x \notin A$ such that $x^p = e$; in that case we saw that the theorem was correct. So we must have $G = (a)$, and G is a cyclic group. This finishes the induction and proves the theorem. $\square$

We are now able to prove the very basic and important

Theorem 2.10.3 (The Fundamental Theorem on Finite Abelian Groups). A finite abelian group is the direct product of cyclic groups.

Proof. Let G be a finite abelian group and p a prime that divides $|G|$. By the Corollary to Lemma 2.10.1, $G = P \times T$, where $|P| = p^n$. By Theorem 2.10.2, $P = A_1 \times A_2 \times \cdots \times A_k$, where the A_i are cyclic subgroups of P. Arguing by induction on $|G|$, we may thus assume that $T = T_1 \times T_2 \times \cdots \times T_q$, where the T_i are cyclic subgroups of T. Thus

$$G = (A_1 \times A_2 \times \cdots \times A_k) \times (T_1 \times T_2 \times \cdots \times T_k)$$

$$= A_1 \times A_2 \times \cdots \times A_k \times T_1 \times T_2 \times \cdots \times T_k. \quad \text{(Prove!)}$$

This very important theorem is now proved. $\square$

We return to abelian groups G of order p^n. We now have at hand that $G = A_1 \times A_2 \times \cdots \times A_k$, where the A_i are cyclic groups of order p^{n_i}. We can arrange the numbering so that $n_1 \ge n_2 \ge \cdots \ge n_k$. Also, $|G| = |A_1 \times A_2 \times \cdots \times A_k| = |A_1| |A_2| \cdots |A_k|$, which gives us that

$$p^n = p^{n_1} p^{n_2} \cdots p^{n_k} = p^{n_1 + n_2 + \cdots + n_k},$$

hence $n = n_1 + n_2 + \cdots + n_k$. Thus the integers $n_i \geq 0$ give us a *partition* of n. It can be shown that these integers $n_1, n_2, \ldots, n_k$—which are called the *invariants* of G—are *unique*. In other words, two abelian groups of order p^n are isomorphic if and only if they have the same invariants. Granted this, it follows that the number of nonisomorphic abelian groups of order p^n is equal to the number of partitions of n.

For example, if $n = 3$, it has the following three partitions: $3 = 3$, $3 = 2 + 1$, $3 = 1 + 1 + 1$, so there are three nonisomorphic abelian groups of order p^3 (independent of p). The groups corresponding to these partitions are a cyclic group of order p^3, the direct product of a cyclic group of order p^2 by one of order p, and the direct product of three cyclic groups of order p, respectively.

For $n = 4$ we see the partitions are $4 = 4$, $4 = 3 + 1$, $4 = 2 + 2$, $4 = 2 + 1 + 1$, $4 = 1 + 1 + 1 + 1$, which are five in number. Thus there are five nonisomorphic groups of order p^4. Can you describe them via the partitions of 4?

Given an abelian group of order $n = p_1^{a_1} p_2^{a_2} \cdots p_k^{a_k}$, where the p_i are distinct primes and the a_i are all positive, then G is the direct product of its so called p_i- Sylow subgroups (see, e.g., the Corollary to Lemma 2.10.1). For each prime p_i there are many groups of order $p_i^{a_i}$ as there are partitions of a_i. So the number of nonisomorphic abelian groups of order $n = p_1^{a_1} \cdots p_k^{a_k}$ is $f(a_1)f(a_2) \cdots f(a_k)$, where $f(m)$ denotes the number of partitions of m. Thus we know how many nonisomorphic finite abelian groups there are for any given order.

For instance, how many nonisomorphic abelian groups are there of order 144? Since $144 = 2^4 3^2$, and there are five partitions of 4, two partitions of 2, there are 10 nonisomorphic abelian groups of order 144.

The material treated in this section has been hard, the path somewhat tortuous, and the effort to understand quite intense. We spare the reader any further agony and so assign no problems to this section.

11.   CONJUGACY AND SYLOW'S THEOREM (OPTIONAL)

In discussing equivalence relations in Section 4 we discussed, as an example of such a relation in a group G, the notion of *conjugacy*. Recall that the element b in G is said to be *conjugate* to $a \in G$ (or merely, a conjugate of a) if there exists an $x \in G$ such that $b = x^{-1}ax$. We showed in Section 4 that this defines an equivalence relation on G. The equivalence class of a, which we denote by cl(a), is called the *conjugacy class* of a.

For a finite group an immediate question presents itself: How large is cl(a)? Of course, this depends strongly on the element a. For instance, if $a \in Z(G)$, the center of G, then $ax = xa$ for all $x \in G$, hence $x^{-1}ax = a$; in other words, the conjugacy class of a in this case consists merely of the element a itself. On the other hand, if cl(a) consists only of the element a, then $x^{-1}ax = a$ for all $x \in G$; this gives us that $xa = ax$ for all $x \in G$, hence $a \in Z(G)$. So $Z(G)$ is characterized as the set of those elements, a, in G whose conjugacy class has only one element, a itself.

For an abelian group G, since $G = Z(G)$, two elements are conjugate if and only if they are equal. So conjugacy is not an interesting relation for abelian groups; however, for nonabelian groups it is a highly interesting notion.

Given $a \in G$, cl(a) consists of all $x^{-1}ax$ as x runs over G. So to determine which are the distinct conjugates of a, we need to know when two conjugates of a coincide, which is the same as asking: When is $x^{-1}ax = y^{-1}ay$? In this case, transposing, we obtain $a(xy^{-1}) = (xy^{-1})a$; in other words, xy^{-1} must commute with a. This brings us to a concept introduced as Example 10 in Section 3, *that of the centralizer of a in G.* We repeat something we did there.

Definition. If $a \in G$, then $C(a)$, the *centralizer of a in G*, is defined by $C(a) = \{ x \in G | xa = ax \}$.

When $C(a)$ arose in Section 3 we showed that it was a subgroup of G. We record this now more officially as

Lemma 2.11.1. For $a \in G$, $C(a)$ is a subgroup of G.

As we saw above, the two conjugates $x^{-1}ax$ and $y^{-1}ay$ of a are equal only if $xy^{-1} \in C(a)$, that is, only if x and y are in the same left coset of $C(a)$ in G. On the other hand, if x and y are in the same left coset of $C(a)$ in G, then $xy^{-1} \in C(a)$, hence $xy^{-1}a = axy^{-1}$. This yields that $x^{-1}ax = y^{-1}ay$. So x and y give rise to the same conjugate of a if and only if x and y are in the same left coset of $C(a)$ in G. *Thus there are as many conjugates of a in G as there are left cosets of $C(a)$ in G.* This is most interesting when G is a finite group, for in that case the number of left cosets of $C(a)$ in G is what we called the *index*, $i_G(C(a))$, of $C(a)$ in G, and is equal to $|G|/|C(a)|$.

We have proved

> **Theorem 2.11.2.** Let G be a finite group and $a \in G$; then the number of distinct conjugates of a in G equals the index of $C(a)$ in G.

In other words, the numbers of elements in $\operatorname{cl}(a)$ equals $i_G(C(a)) = |G|/|C(a)|$.

This theorem, although it was relatively easy to prove, is very important and has many consequences. We shall see a few of these here.

One such consequence is a kind of bookkeeping result. Since conjugacy is an equivalence relation on G, G is the union of the disjoint conjugacy classes. Moreover, by Theorem 2.11.2, we know how many elements there are in each class. Putting all this information together, we get

> **Theorem 2.11.3 (The Class Equation).** If G is a finite group, then
>
> $$|G| = \sum_a i_G(C(a)) = \sum_a \frac{|G|}{|C(a)|},$$
>
> where the sum runs over one a from each conjugacy class.

It is almost a sacred tradition among mathematicians to give, as the first application of the class equation, a particular theorem about groups of order p^n, where p is a prime. Not wanting to be accused of heresy, we follow this tradition and prove the pretty and important

> **Theorem 2.11.4.** If G is a group of order p^n, where p is a prime, then $Z(G)$, the center of G, is not trivial (i.e., there exists an element $a \neq e$ in G such that $ax = xa$ for all $x \in G$).

Proof. We shall exploit the class equation to carry out the proof. Let $z = |Z(G)|$; as we pointed out previously, z is then the number of elements in G whose conjugacy class has only one element. Since $e \in Z(G)$, $z \geq 1$. For any element b outside $Z(G)$, its conjugacy class contains more than one element and $|C(b)| < |G|$. Moreover, since $|C(b)|$ divides $|G|$ by Lagrange's theorem, $|C(b)| = p^{n(b)}$, where $1 \leq n(b) < n$. We divide the pieces of the class equation into two parts: that coming from the center, and the rest. We get, this way,

$$p^n = |G| = z + \sum_{b \notin Z(G)} \frac{|G|}{|C(b)|} = z + \sum_{n(b) < n} \frac{p^n}{p^{n(b)}} = z + \sum_{n(b) < n} p^{n - n(b)}.$$

Clearly, p divides the left-hand side, p^n, and divides $\sum_{n(b) < n} p^{n - n(b)}$. The net result of this is that $p|z$, and since $z \geq 1$, we have that z is *at least p*. So since $z = |Z(G)|$, there must be an element $a \neq e$ in $Z(G)$, which proves the theorem. $\square$

This last theorem has an interesting application, which some readers may have seen in solving Problem 45 of Section 5. This is

Theorem 2.11.5. If G is a group of order p^2, where p is a prime, then G is abelian.

Proof. By Theorem 2.11.4, $Z(G) \neq (e)$, so that there is an element, a, of order p in $Z(G)$. If $A = (a)$, the subgroup generated by a, then $A \subset Z(G)$, hence $A \subset C(x)$ for all $x \in G$. Given $x \in G$, $x \notin A$, then $C(x) \supset A$ and $x \in C(x)$; so $|C(x)| > p$, yet $|C(x)|$ must divide p^2. The net result of this is that $|C(x)| = p^2$, so $C(x) = G$, whence $x \in Z(G)$. Since every element of G is in the center of G, G must be abelian. $\square$

In the problems to come we shall give many applications of the nature of groups of order p^n, where p is a prime. *The natural attack on virtually all these problems follows the lines of the argument we are about to give.* We choose one of a wide possible set of choices to illustrate this technique.

Theorem 2.11.6. If G is a group of order p^n, p a prime, then G contains a normal subgroup of order p^{n-1}.

Proof. We proceed by induction on n. If $n = 1$, then G is of order p and (e) is the required normal subgroup of order $p^{1-1} = p^0 = 1$.

Suppose that we know that for some k, every group of order p^k has a normal subgroup of order p^{k-1}. Let G be of order p^{k+1}; by Theorem 2.11.4 there exists an element a of order p in $Z(G)$, the center of G. Thus the subgroup $A = (a)$ generated by a is of order p and is normal in G. Consider $\Gamma = G/A$; Γ is a group of order $|G|/|A| = p^{k+1}/p = p^k$ by Theorem 2.6.3. Since Γ has order p^k, we know that Γ has a normal subgroup M of order p^{k-1}. Since Γ is a homomorphic image of G, by the Second Homomorphism Theorem (Theorem 2.7.2) there is a normal subgroup N in G, $N \supset A$, such that $N/A = M$. But then we have

$$|M| = | N / A | = \frac{|N|}{|A|},$$

that is, $p^{k-1} = |N|/p$, leading us to $|N| = p^k$. N is our required normal subgroup in G of order p^k. This completes the induction and so proves the theorem. $\square$

By far the most important application we make of the class equation is the proof of a far-reaching theorem due to Sylow, a Norwegian mathematician, who proved it in 1871. We already showed this theorem to be true for abelian groups. We shall now prove it for any finite group. It is impossible to overstate the importance of *Sylow's Theorem* in the study of finite groups. Without it the subject would not get off the ground.

Theorem 2.11.7 (Sylow's Theorem). Suppose that G is a group of order $p^n m$, where p is a prime and $p \nmid m$. Then G has a subgroup of order p^n.

Proof. If $n = 0$, then $p \nmid |G|$ and there is nothing to prove. We therefore assume that $n \geq 1$. Here, again, we proceed by induction on $|G|$, assuming the result to be true for all groups H such that $|H| < |G|$.

Suppose that the result is false for G. Then, by our induction hypothesis, p^n *cannot divide* $|H|$ for any subgroup H of G if $H \neq G$. *In particular, if* $a \notin Z(G)$, *then* $C(a) \neq G$, *hence* $p^n \nmid |C(a)|$. Thus $p \mid |G|/|C(a)| = i_G(C(a))$ for $a \notin Z(G)$.

Write down the class equation for G following the lines of the argument in Theorem 2.11.4. If $z = |Z(G)|$, then $z \geq 1$ and

$$p^n m = |G| = z + \sum_{a \notin Z(G)} i_G(C(a)).$$

But $p \mid i_G(C(a))$ if $a \notin Z(G)$, so $p \mid \sum_{a \notin Z(G)} i_G(C(a))$; since $p \mid p^n m$, we get $p \mid z$. By Cauchy's Theorem there is an element a of order p in $Z(G)$, since $p \mid z = |Z(G)|$. If A is the subgroup generated by a, then $|A| = p$ and $A \lhd G$, since $a \in Z(G)$. Consider $\Gamma = G/A$; $|\Gamma| = |G|/|A| = p^n m/p = p^{n-1} m$. Since $|\Gamma| < |G|$, by our induction hypothesis Γ has a subgroup M of order p^{n-1}. However, by the Second Homomorphism Theorem there is a subgroup P of G such that $P \supset A$ and $P/A = M$. Therefore, $|P| = |M| |A| = p^{n-1} p = p^n$; P is thus the sought-after subgroup of G of order p^n, contradicting that we assumed that G had no such subgroup. This completes the induction, and Sylow's Theorem is established. $\square$

Actually, Sylow's Theorem consists of three parts, of which we only proved the first. The other two are (assuming $p^n m = |G|$, where $p \nmid m$):

1. Any two subgroups of order p^n in G are conjugate; that is, if $|P| = |Q| = p^n$, P, Q subgroups of G, then for some $x \in G$, $Q = x^{-1} P x$.
2. The number of subgroups of order p^n in G is of the form $1 + kp$ and divides $|G|$.

Since these subgroups of order p^n pop up all over the place, they are called *p-Sylow subgroups* of G. An abelian group has one p-Sylow subgroup for every prime p dividing its order. This is far from true in the general case. For instance, if $G = S_3$, the symmetric group of degree 3, which has order $6 = 2 \cdot 3$, there are three 2-Sylow subgroups (of order 2) and one 3-Sylow subgroup (of order 3).

For those who want to see several proofs of that part of Sylow's Theorem which we proved above, and of the other two parts, they might look at the appropriate section of our book *Topics in Algebra*.

Problems

Easier Problems

1. In S_3, the symmetric group of degree 3, find all the conjugacy classes, and check the validity of the class equation by determining the orders of the centralizers of the elements of S_3.

2. Do Problem 1 for G the dihedral group of order 8.

3. If $a \in G$, show that $C(x^{-1}ax) = x^{-1}C(a)x$.

4. If φ is an automorphism of G, show that $C(\varphi(a)) = \varphi(C(a))$ for $a \in G$.

5. If $|G| = p^3$ and $|Z(G)| \geq p^2$, prove that G is abelian.

6. If P is a p-Sylow subgroup of G and $P \triangleleft G$, prove that P is the only p-Sylow subgroup of G.

7. If $P \triangleleft G$, P a p-Sylow subgroup of G, prove that $\varphi(P) = P$ for every automorphism φ of G.

8. Use the class equation to give a proof of Cauchy's Theorem.

   If H is a subgroup of G, let $N(H) = \{ x \in G | x^{-1}Hx = H \}$. This *does not mean* that $x \in N(H)$, $a \in H$, then $xa = ax$. For instance, if $H \triangleleft G$, then $N(H) = G$, yet H need not be in the center of G.

9. Prove that $N(H)$ is a subgroup of G, $H \subset N(H)$ and $H \triangleleft N(H)$.

10. Prove that $N(x^{-1}Hx) = x^{-1}N(H)x$.

11. If P is a p-Sylow subgroup of G, prove that P is a p-Sylow subgroup of $N(P)$ and is the only p-Sylow subgroup of $N(P)$.

12. If P is a p-Sylow subgroup and $a \in G$ is of order p^m for some m, show that if $a^{-1}Pa = P$ then $a \in P$.

13. Prove that if G is a finite group and H is a subgroup of G, then the number of distinct subgroups $x^{-1}Hx$ of G equals $i_G(N(H))$.

14. If P is a p-Sylow subgroup of G, show that the number of distinct $x^{-1}Px$ *cannot* be a multiple of p.

15. If $N \triangleleft G$, let $B(N) = \{ x \in G | xa = ax$ for all $a \in N \}$. Prove that $B(N) \triangleleft G$.

Middle-Level Problems

16. Show that a group of order 36 has a normal subgroup of order 3 or 9. (**Hint:** See Problem 40 of Section 5.)

17. Show that a group of order 108 has a normal subgroup of order 9 or 27.

18. If P is a p-Sylow subgroup of G, show that $N(N(P)) = N(P)$.

19. If $|G| = p^n$, show that G has a subgroup of order p^m for all $1 \leq m \leq n$.

20. If $p^m | |G|$, show that G has a subgroup of order p^m.

21. If $|G| = p^n$ and $H \neq G$ is a subgroup of H, show that $N(H) \supsetneq H$.

22. Show that any subgroup of order p^{n-1} in a group G of order p^n is normal in G.

Harder Problems

23. Let G be a group, H a subgroup of G. Define for $a, b \in G$, $a \sim b$ if $b = h^{-1}ah$ for some $h \in H$. Prove that
   (a) this defines an equivalence relation on G.
   (b) If $\{a\}$ is the equivalence class of a, show that if G is a finite group, then $\{a\}$ has m elements where m is the index of $H \cap C(a)$ in H.

24. If G is a group, H a subgroup of G, define for A, B subgroups that $B \sim A$ if $B = h^{-1}Ah$ for some $h \in H$.
   (a) Prove that this defines an equivalence relation on the set of subgroups of G.
   (b) If G is finite, show that the number of distinct subgroups equivalent to A equals the index of $N(A) \cap H$ in H.

25. If P is a p-Sylow subgroup of G, let S be the set of all p-Sylow subgroups of G. For $Q_1, Q_2 \in S$ define $Q_1 \sim Q_2$ if $Q_2 = a^{-1}Q_1a$ with $a \in P$. Prove, using this relation, that if $Q \neq P$, then the number of distinct $a^{-1}Qa$, with $a \in P$, is a multiple of p.

26. Using the result of Problem 25, show that the number of p-Sylow subgroups of G is of the form $1 + kp$. (This is the third part of Sylow's Theorem.)

27. Let P be a p-Sylow subgroup of G, and Q another one. Suppose that $Q \neq x^{-1}Px$ for any $x \in G$. Let S be the set of all $y^{-1}Qy$, as y runs over G. For $Q_1, Q_2 \in S$ define $Q_1 \sim Q_2$ if $Q_2 = a^{-1}Q_1 a$, where $a \in P$.

    (a) Show that this implies that the number of distinct $y^{-1}Qy$ is a multiple of p.

    (b) Using the result of Problem 14, show that the result of Part (a) cannot hold.

    (c) Prove from this that given any two p-Sylow subgroups P and Q of G, then $Q = x^{-1}Px$ for some $x \in G$.

    (This is the second part of Sylow's Theorem.)

28. If H is a subgroup of G of order p^m show that H is contained in some p-Sylow subgroup of G.

29. If P is a p-Sylow subgroup of G and $a, b \in Z(P)$ are conjugate in G, prove that they are already conjugate in $N(P)$.

CHAPTER 3

The Symmetric Group

1. PRELIMINARIES

Let us recall a theorem proved in Chapter 2 for abstract groups. This result, known as *Cayley's Theorem* (Theorem 2.5.1), asserts that any group G is isomorphic to a subgroup of $A(S)$, the set of 1-1 mappings of the set S onto itself, for some suitable S. In fact, in the proof we gave we used for S the group G itself viewed merely as a set.

Historically, groups arose this way first, long before the notion of an abstract group was defined. We find in the work of Lagrange, Abel, Galois, and others, results on groups of permutations proved in the late eighteenth and early nineteenth centuries. Yet is was not until the mid-nineteenth century that Cayley more or less introduced the abstract concept of a group.

Since the structure of isomorphic groups is the same, Cayley's Theorem points out a certain universal character for the groups $A(S)$. If we knew the structure of all subgroups of $A(S)$ for any set S, we would know the structure of all groups. This is much too much to expect. Nevertheless, one could try to exploit this embedding of an arbitrary group G isomorphcially into some $A(S)$. This has the advantage of transforming G as an abstract system into something more concrete, namely as a set of nice mappings of some set onto itself.

We shall not be concerned with the subgroups of $A(S)$ for an arbitrary set S. If S is infinite, $A(S)$ is a very wild and complicated object. Even if S is finite, the complete nature of $A(S)$ is virtually impossible to determine.

In this chapter we consider only $A(S)$ for S a finite set. Recall that if S has n elements, then we call $A(S)$ the *symmetric group of degree n*, and denote it by S_n. The elements of S_n are called *permutations*; we shall denote them by lowercase Greek letters.

Since we multiplied two elements $\sigma, \tau \in A(S)$ by the rule $(\sigma\tau)(s) = \sigma(\tau(s))$ this will have the effect that when we introduce the appropriate symbols to represent the elements of S_n, these symbols, or permutations, will multiply from *right to left*. If the readers look at some other book on algebra, they should make sure which way the permutations are being multiplied: right to left or left to right. Very often, algebraists multiply permutations from *left to right*. To be consistent with our definition of the composition of elements in S_n, we do it from right to left.

By Cayley's Theorem we know that if G is a finite group of order n, then G is isomorphic to a subgroup of S_n and S_n has $n!$ elements. Speaking loosely, we usually say that G is a subgroup of S_n. Since n is so much smaller than $n!$ for n even modestly large, our group occupies only a tiny little corner in S_n. It would be desirable to embed G in an S_n for n as small as possible. For certain classes of finite groups this is achievable.

Let S be a finite set having n elements; we might as well suppose that $S = \{x_1, x_2, \ldots, x_n\}$. Given the permutation $\sigma \in S_n = A(S)$, then $\sigma(x_k) \in S$ for $k = 1, 2, \ldots, n$, so $\sigma(x_k) = x_{i_k}$ for some i_k, $1 \le i_k \le n$. Because σ is 1-1, if $j \ne k$, then $x_{i_j} = \sigma(x_j) \ne \sigma(x_k) = x_{i_k}$; therefore, the numbers $i_1, i_2, \ldots, i_n$ are merely the numbers $1, 2, \ldots, n$ shuffled about in some order.

Clearly, the action of σ on S is determined by what σ does to the subscript j of x_j, so the symbol "x" is really excess baggage and, as such, can be discarded. In short, we may assume that $S = \{1, 2, \ldots, n\}$.

Let's recall what is meant by the *product* of two elements of $A(S)$. If $\sigma, \tau \in A(S)$, then we defined $\sigma\tau$ by $(\sigma\tau)(s) = \sigma(\tau(s))$ for every $s \in S$. We showed in Section 4 of Chapter 1 that $A(S)$ satisfied four properties that we used later as the model to define the notion of an abstract group. Thus S_n, in particular, is a group relative to the product of mappings.

Our first need is some handy way of denoting a permutation, that is, an element σ in S_n. One clear way is to make a table of what σ does to each element of S. This might be called the *graph* of σ. We did this earlier, writing out σ, say $\sigma \in S_3$, in the fashion: $\sigma : x_1 \rightarrow x_2$, $x_2 \rightarrow x_3$, $x_3 \rightarrow x_1$. But this is cumbersome and space consuming. We certainly can make it more compact by dropping the x's and writing $\sigma = \begin{pmatrix} 1 & 2 & 3 \\ 2 & 3 & 1 \end{pmatrix}$. In this symbol the number in the second row is the image under of the number in the first row directly above it. There is nothing holy about 3 in all this; it works equally well for any n.

If $\sigma \in S_n$ and $\sigma(1) = i_1$, $\sigma(2) = i_2, \ldots, \sigma(n) = i_n$, we use the symbol $\begin{pmatrix} 1 & 2 & \cdots & n \\ i_1 & i_2 & \cdots & i_n \end{pmatrix}$ to represent σ and we write $\sigma = \begin{pmatrix} 1 & 2 & \cdots & n \\ i_1 & i_2 & \cdots & i_n \end{pmatrix}$. Note that it is not necessary to write the first row in the usual order 1 2 $\cdots$ n; any way we write the first row, as long as we carry the i_j's along accordingly we still have σ. For instance, in the example in S_3 cited,

$$\sigma = \begin{pmatrix} 1 & 2 & 3 \\ 2 & 3 & 1 \end{pmatrix} = \begin{pmatrix} 3 & 1 & 2 \\ 1 & 2 & 3 \end{pmatrix} = \begin{pmatrix} 2 & 1 & 3 \\ 3 & 2 & 1 \end{pmatrix}.$$

If we know $\sigma = \begin{pmatrix} 1 & 2 & \cdots & n \\ i_1 & i_2 & \cdots & i_n \end{pmatrix}$, what is σ^{-1}? It is easy, just flip the symbol for σ over to get $\sigma^{-1} = \begin{pmatrix} i_1 & i_2 & \cdots & i_n \\ 1 & 2 & \cdots & n \end{pmatrix}$. (Prove!) In our example

$$\sigma = \begin{pmatrix} 1 & 2 & 3 \\ 2 & 3 & 1 \end{pmatrix}, \qquad \sigma^{-1} = \begin{pmatrix} 2 & 3 & 1 \\ 1 & 2 & 3 \end{pmatrix} = \begin{pmatrix} 1 & 2 & 3 \\ 3 & 1 & 2 \end{pmatrix}.$$

The identity element—which we shall write as e—is merely $e = \begin{pmatrix} 1 & 2 & \cdots & n \\ 1 & 2 & \cdots & n \end{pmatrix}$.

How does the product in S_n translate in terms of these symbols? Since $\sigma\tau$ means: "First apply τ and to the result of this apply σ," in forming the product of the symbols for σ and τ we look at the number k in the first row of τ and see what number i_k is directly below k in the second row of τ. We then look at the spot i_k in the first row of σ and see what is directly below it in the second row of σ. This is the image of

k under $\sigma\tau$. We then run through $k = 1, 2, \ldots, n$ and get the symbol for $\sigma\tau$. We just do this visually.

We illustrate this with two permutations

$$\sigma = \begin{pmatrix} 1 & 2 & 3 & 4 & 5 \\ 2 & 3 & 1 & 5 & 4 \end{pmatrix} \text{ and } \tau = \begin{pmatrix} 1 & 2 & 3 & 4 & 5 \\ 3 & 4 & 5 & 1 & 2 \end{pmatrix}$$

is S_5. Then $\sigma\tau = \begin{pmatrix} 1 & 2 & 3 & 4 & 5 \\ 1 & 5 & 4 & 2 & 3 \end{pmatrix}$

Even the economy achieved this way is not enough. After all, the first row is always $1 \ 2 \ \cdots \ n$, so we could dispense with it, and write $\sigma = \begin{pmatrix} 1 & 2 & \cdots & n \\ i_1 & i_2 & \cdots & i_n \end{pmatrix}$ as $(i_1, i_2, \ldots, i_n)$. This is fine, but in the next section we shall find a better and briefer way of representing permutations.

Problems

1. Find the products.

   (a) $\begin{pmatrix} 1 & 2 & 3 & 4 & 5 & 6 \\ 6 & 4 & 5 & 2 & 1 & 3 \end{pmatrix}\begin{pmatrix} 1 & 2 & 3 & 4 & 5 & 6 \\ 2 & 3 & 4 & 5 & 6 & 1 \end{pmatrix}$.

   (b) $\begin{pmatrix} 1 & 2 & 3 & 4 & 5 \\ 2 & 1 & 3 & 4 & 5 \end{pmatrix}\begin{pmatrix} 1 & 2 & 3 & 4 & 5 \\ 3 & 2 & 1 & 4 & 5 \end{pmatrix}$

   (c) $\begin{pmatrix} 1 & 2 & 3 & 4 & 5 \\ 4 & 1 & 3 & 2 & 5 \end{pmatrix}^{-1}\begin{pmatrix} 1 & 2 & 3 & 4 & 5 \\ 2 & 1 & 3 & 4 & 5 \end{pmatrix}$

   $\times \begin{pmatrix} 1 & 2 & 3 & 4 & 5 \\ 4 & 1 & 3 & 2 & 5 \end{pmatrix}$.

2. Evaluate all the powers of each permutation (i.e., find σ^k for all k).

   (a) $\begin{pmatrix} 1 & 2 & 3 & 4 & 5 & 6 \\ 2 & 3 & 4 & 5 & 6 & 1 \end{pmatrix}$.

   (b) $\begin{pmatrix} 1 & 2 & 3 & 4 & 5 & 6 & 7 \\ 2 & 1 & 3 & 4 & 6 & 5 & 7 \end{pmatrix}$.

   (c) $\begin{pmatrix} 1 & 2 & 3 & 4 & 5 & 6 \\ 6 & 4 & 5 & 2 & 1 & 3 \end{pmatrix}$.

3. Prove that $\begin{pmatrix} 1 & 2 & \cdots & n \\ i_1 & i_2 & \cdots & i_n \end{pmatrix}^{-1} = \begin{pmatrix} i_1 & i_2 & \cdots & i_n \\ 1 & 2 & \cdots & n \end{pmatrix}$.

4. Find the order of each element in Problem 2.

5. Find the order of the products you obtained in Problem 1.

2. CYCLE DECOMPOSITION

We continue the process of simplifying the notation used to represent a given permutation. In doing so, we get something more than just a new symbol;-we get a device to decompose any permutation as a product of particularly nice permutations.

> **Definition.** Let $i_1, i_2, \ldots, i_k$ be k distinct integers in $S = \{1, 2, \ldots, n\}$. The symbol $(i_1 \ \ i_2 \ \ \cdots \ \ i_k)$ will represent the permutation $\sigma \in S_n$, where $\sigma(i_1) = i_2, \sigma(i_2) = i_3, \ldots, \sigma(i_j) = i_{j+1}$ for $j < k$, $\sigma(i_k) = i_1$, and $\sigma(s) = s$ for any $s \in S$ if $s \neq i_1, i_2, \ldots, i_k$.

Thus, in S_7, the permutation $(1 \ \ 3 \ \ 5 \ \ 4)$ is the permutation $\begin{pmatrix} 1 & 2 & 3 & 4 & 5 & 6 & 7 \\ 3 & 2 & 5 & 1 & 4 & 6 & 7 \end{pmatrix}$. We call a permutation of the form $(i_1 \ \ i_2 \ \ \cdots \ \ i_k)$ a k-*cycle*. For the special case $k = 2$, the permutation $(i_1 \ \ i_2)$ is called a *transposition*. Note that if $\sigma = (i_1 \ \ i_2 \ \ \cdots \ \ i_k)$, then σ is likewise $(i_k \ \ i_1 \ \ i_2 \ \ \cdots \ \ i_{k-1})$, $(i_{k-1} \ \ i_k \ \ i_1 \ \ i_2 \ \ \cdots \ \ i_{k-2})$, and so on. (Prove!) For example,

$$(1 \ \ 3 \ \ 5 \ \ 4) = (4 \ \ 1 \ \ 3 \ \ 5) = (5 \ \ 4 \ \ 1 \ \ 3) = (3 \ \ 5 \ \ 4 \ \ 1).$$

Given two cycles, say a k-cycle and an m-cycle, they are said to be *disjoint cycles* if they have no integer in common. Whence $(1 \ \ 3 \ \ 5)$ and $(4 \ \ 2 \ \ 6 \ \ 7)$ in S_7 are disjoint cycles.

Given two disjoint cycles in S_n, we claim that they commute. We leave the proof of this to the reader, with the suggestion that if σ, τ are disjoint cycles, the reader should verify that $(\sigma\tau)(i) = (\tau\sigma)(i)$ for every $i \in S = \{1, 2, \ldots, n\}$. We state this result as

Lemma 3.2.1. If $\sigma, \tau \in S_n$ are disjoint cycles, then $\sigma\tau = \tau\sigma$.

Let's consider a particular k-cycle $\sigma = (1 \;\; 2 \;\; \cdots \;\; k)$ in S_n. Clearly, $\sigma(1) = 2$ by the definition given above; how is 3 related to 1? Since $\sigma(2) = 3$, we have $\sigma^2(1) = \sigma(2) = 3$. Continuing, we see that $\sigma^j(1) = j + 1$ for $j \le k - 1$, while $\sigma^k(1) = 1$. In fact, we see that $\sigma^k = e$, where e is the identity element in S_n.

There are two things to be concluded from the paragraph above.

1. The order of a k-cycle, as an element of S_n, is k. (Prove!)
2. If $\sigma = (i_1 \;\; i_2 \;\; \cdots \;\; i_k)$ is a k-cycle, then the *orbit* of i_1 under σ (see Problem 27 in Section 4 of Chapter 1) is $\{i_1, i_2, \ldots, i_k\}$. So we can see that the k-cycle $\sigma = (i_1 \;\; i_2 \;\; \cdots \;\; i_k)$ is

$$\sigma = \left(i_1 \;\; \sigma(i_1) \;\; \sigma^2(i_1) \;\; \cdots \;\; \sigma^{k-1}(i_1)\right).$$

Given *any* permutation τ in S_n for $i \in \{1, 2, \ldots, n\}$, consider the *orbit* of i under τ; we have that this orbit is $\{i, \tau(i), \tau^2(i), \ldots, \tau^{s-1}(i)\}$, where $\tau^s(i) = i$ and s is the smallest positive integer with this property. Consider the s-cycle $(i \;\; \tau(i) \;\; \tau^2(i) \;\; \cdots \;\; \tau^{s-1}(i))$; we call it the *cycle of τ determined by i.*

We take a specific example and find all its cycles. Let

$$\tau = \begin{pmatrix} 1 & 2 & 3 & 4 & 5 & 6 & 7 & 8 & 9 \\ 3 & 9 & 4 & 1 & 5 & 6 & 2 & 7 & 8 \end{pmatrix};$$

what is the cycle of τ determined by 1? We claim that it is $(1 \;\; 3 \;\; 4)$. Why? τ takes 1 into 3, 3 into 4 and 4 into 1, and since $\tau(1) = 3$, $\tau^2(1) = \tau(3) = 4$, $\tau^3(1) = \tau(4) = 1$. We can get this visually by weaving through

$$\begin{pmatrix} 1 & 2 & 3 & 4 & 5 & 6 & 7 & 8 & 9 \\ 3 & 9 & 4 & 1 & 5 & 6 & 2 & 7 & 8 \end{pmatrix}$$

with the dotted dashed line. What is the cycle of τ determined by 2? Weaving through

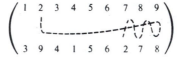

with the dashed line, we see that the cycle of τ determined by 2 is $(2 \;\; 9 \;\; 8 \;\; 7)$. The cycle of τ determined by 5 and 6 are (5) and (6),

respectively, since 5 and 6 are left fixed by τ. So the cycles of τ are
(1 3 4), (2 9 8 7), (5), and (6). Therefore we have that $\tau =$
(1 3 4)(2 9 8 7)(5)(6), where we view these cycles—as defined
above—as permutations in S_9 because every integer in $S = \{1, 2, \ldots, 9\}$
appears in one, and only one, cycle and the image of any i under τ is
read off from the cycle in which it appears.

There is nothing special about the permutation τ above that made the
argument we gave go through. The same argument would hold for *any*
permutation in S_n for *any* n. We leave the formal writing down of the
proof to the reader.

Theorem 3.2.2. Every permutation in S_n is the product of
disjoint cycles.

In writing a permutation σ as a product of disjoint cycles, we omit
all 1-cycles; that is, we ignore the i's such that $\sigma(i) = i$. Thus $\sigma =$
(1 2 3)(4 5) in S_7 is the way that we would write $\sigma =$
(1 2 3)(4 5)(6)(7). In other words, writing σ as a product of
k-cycles, with $k > 1$, we assume that σ leaves fixed any integer not
present in any of the cycles. Thus in the group S_{11} the permutation
$\tau = (1 5 6)(2 3 9 8 7)$ leaves fixed 4, 10, and 11.

Lemma 3.2.3. If τ in S_n is a k-cycle, then the order of τ is
k; that is, $\tau^k = e$ and $\tau^j \neq e$ for $0 < j < k$.

Consider the permutation $\tau = (1 2)(3 4 5 6)(7 8 9)$ in S_9.
What is its order? Since the disjoint cycles (1 2), (3 4 5 6),
(7 8 9) commute, $\tau^m = (1 2)^m(3 4 5 6)^m(7 8 9)^m$; in order
that $\tau^m = e$ we need (1 2)$^m = e$, (3 4 5 6)$^m = e$, (7 8 9)$^m = e$.
(Prove!) To have (7 8 9)$^m = e$, we must have $3|m$, since (7 8 9) is
of order 3; to have (3 4 5 6)$^m = e$, we must have $4|m$, because
(3 4 5 6) is of order 4, and to have (1 2)$^m = e$, we must have $2|m$,
because (1 2) is of order 2. This tells us that m must be divisible by 12.
On the other hand,

$$\tau^{12} = (1 2)^{12}(3 4 5 6)^{12}(7 8 9)^{12} = e.$$

So τ is of order 12.

Here, again, the special properties of τ do not enter the picture. What we did for τ works for any permutation. To formulate this properly, recall that the least common multiple of m and n is the smallest positive integer v which is divisible by m and by n. (See Problem 7, Chapter 1, Section 5). Then we have

> **Theorem 3.2.4.** Let $\sigma \in S_n$ have its cycle decomposition into *disjoint* cycles of length $m_1, m_2, \ldots, m_k$. Then the order of σ is the least common multiple of $m_1, m_2, \ldots, m_k$.

Proof. Let $\sigma = \tau_1 \tau_2 \cdots \tau_k$, where the τ_i are disjoint cycles of length m_i. Since the τ_i are disjoint cycles, $\tau_i \tau_j = \tau_j \tau_i$; therefore if M is the least common multiple of $m_1, m_2, \ldots, m_k$, then $\sigma^M = (\tau_1 \tau_2 \cdots \tau_k)^M = \tau_1^M \tau_2^M \cdots \tau_k^M = e$ (since $\tau_i^M = e$ because τ_i is of order m_i and $m_i | M$). Therefore, the order of σ is *at most* M. On the other hand, if $\sigma^N = e$, then $\tau_1^N \tau_2^N \cdots \tau_k^N = e$; this forces each $\tau_i^N = e$, (prove!) because τ_i are disjoint permutations, so $m_i | N$, since τ_i is of order m_i. Thus N is divisible by the least common multiple of $m_1, m_2, \ldots, m_k$, so $M | N$. Consequently, we see that σ is of order M as claimed in the theorem. $\quad\square$

Note that the disjointness of the cycles in the theorem is imperative. For instance, $(1\ 2)$ and $(1\ 3)$, which are not disjoint, are each of order 2, but their product $(1\ 2)(1\ 3) = (1\ 3\ 2)$ is of order 3.

Let's consider Theorem 3.2.4 in the context of a card shuffle. Suppose that we shuffle a deck of 13 cards in such a way that the top card is put into the position of the 3rd card, the second in that of the 4th, ..., the ith into the $i + 2$ position, working mod 13. As a permutation, σ, of $1, 2, \ldots, 13$ the shuffle becomes

$$\sigma = \begin{pmatrix} 1 & 2 & 3 & 4 & 5 & 6 & 7 & 8 & 9 & 10 & 11 & 12 & 13 \\ 3 & 4 & 5 & 6 & 7 & 8 & 9 & 10 & 11 & 12 & 13 & 1 & 2 \end{pmatrix},$$

and σ is merely the 13-cycle $(1\ 3\ 5\ 7\ 9\ 11\ 13\ 2\ 4\ 6\ 8\ 10\ 12)$, so σ is of order 13. How many times must we repeat this shuffle to get the cards back to their original order? The answer is merely the order of σ, that is, 13. So it takes 13 repeats of the shuffle to get the cards back to their original order.

Let's give a twist to the shuffle above. Suppose that we shuffle the cards as follows. First take the top card and put it into the second-to-last place, and then follow it by the shuffle given above. How many repeats are now needed to get the cards back to their original order? The first operation is the shuffle given by the permutation $\tau =$ (1 12 11 10 9 8 7 6 5 4 3 2) followed by σ above. So we must compute $\sigma\tau$ and find its order. But

$$\sigma\tau = (1 \quad 3 \quad 5 \quad 7 \quad 9 \quad 11 \quad 13 \quad 2 \quad 4 \quad 6 \quad 8 \quad 10 \quad 12)$$

$$\times (1 \quad 12 \quad 11 \quad 10 \quad 9 \quad 8 \quad 7 \quad 6 \quad 5 \quad 4 \quad 3 \quad 2)$$

$$= (1)(2 \quad 3 \quad 4 \quad 5 \quad 6 \quad 7 \quad 8 \quad 9 \quad 10 \quad 11 \quad 12 \quad 13),$$

so is of order 12. So it would take 12 repeats of the shuffle to get back to the original order.

Can you find a shuffle of the 13 cards that would require 42 repeats? Or 20 repeats? What shuffle would require the greatest number of repeats, and what would this number be?

We return to the general discussion. Consider the permutation (1 2 3); we see that (1 2 3) = (1 3)(1 2). We can also see that (1 2 3) = (2 3)(1 3). So two things are evident. First, we can write (1 2 3) as the product of two transpositions, and in at least two distinct ways. CGiven the k-cycle $(i_1 \quad i_2 \quad \cdots \quad i_k)$, then $(i_1 \quad i_2 \quad \cdots \quad i_k)$ $= (i_1 \quad i_k)((i_1 \quad i_{k-1}) \cdots (i_1 \quad i_2),$, so every k- cycle is product of $k-1$ transpositions (if $k > 1$) and this can be done in several ways, so not in a unique way. Because every permutation is the product of disjoint cycles and every cycle is a product of transpositions we have

Theorem 3.2.5. Every permutation in S_n is the product of transpositions.

This theorem is really not surprising for it says, after all, nothing more or less than that any permutation can be effected by carrying out a series of interchanges of two objects at a time.

We saw that there is a lack of uniqueness in representing a given permutation as a product of transpositions. But, as we shall see in Section 3, some aspects of this decomposition are indeed unique.

Problems

Easier Problems

1. Show that if σ, τ are two disjoint cycles, then $\sigma\tau = \tau\sigma$.

2. Find the cycle decomposition and order.

(a) $\begin{pmatrix} 1 & 2 & 3 & 4 & 5 & 6 & 7 & 8 & 9 \\ 3 & 1 & 4 & 2 & 7 & 6 & 9 & 8 & 5 \end{pmatrix}$.

(b) $\begin{pmatrix} 1 & 2 & 3 & 4 & 5 & 6 & 7 \\ 7 & 6 & 5 & 4 & 3 & 2 & 1 \end{pmatrix}$.

(c) $\begin{pmatrix} 1 & 2 & 3 & 4 & 5 & 6 & 7 \\ 7 & 6 & 5 & 3 & 4 & 2 & 1 \end{pmatrix}$
$\times \begin{pmatrix} 1 & 2 & 3 & 4 & 5 & 6 & 7 \\ 2 & 3 & 1 & 5 & 6 & 7 & 4 \end{pmatrix}$.

3. Express as the product of disjoint cycles and find the order.
   (a) $(1 \ 2 \ 3 \ 5 \ 7)(2 \ 4 \ 7 \ 6)$.
   (b) $(1 \ 2)(1 \ 3)(1 \ 4)$.
   (c) $(1 \ 2 \ 3 \ 4 \ 5)(1 \ 2 \ 3 \ 4 \ 6)(1 \ 2 \ 3 \ 4 \ 7)$.
   (d) $(1 \ 2 \ 3)(1 \ 3 \ 2)$.
   (e) $(1 \ 2 \ 3)(3 \ 5 \ 7 \ 9)(1 \ 2 \ 3)^{-1}$.
   (f) $(1 \ 2 \ 3 \ 4 \ 5)^3$.

4. Give a complete proof of Theorem 3.2.2.

5. Show that a k-cycle has order k.

6. Find a shuffle of a deck of 13 cards that requires 42 repeats to return the cards to their original order.

7. Do Problem 6 for a shuffle requiring 20 repeats.

8. Express the permutations in Problem 3 as the product of transpositions.

9. Given the two transpositions $(1 \ 2)$ and $(1 \ 3)$, find a permutation σ such that $\sigma(1 \ 2)\sigma^{-1} = (1 \ 3)$.

10. Prove that there is no permutation σ such that $\sigma(1 \quad 2)\sigma^{-1} =$ (1 2 3).

11. Prove that there is a permutation σ such that $\sigma(1 \quad 2 \quad 3)\sigma^{-1} =$ (4 5 6).

12. Prove that there is no permutation σ such that $\sigma(1 \quad 2 \quad 3)\sigma^{-1} =$ (1 2 4)(5 6 7).

Middle-Level Problems

13. Prove that (1 2) cannot be written as the product of disjoint 3-cycles.

14. Prove that for any permutation σ, $\sigma\tau\sigma^{-1}$ is a transposition if τ is a transposition.

15. Show that if τ is a k-cycle, then $\sigma\tau\sigma^{-1}$ is also a k-cycle, for any permutation σ.

16. Let Φ be an automorphism of S_3. Show that there is an element $\sigma \in S_3$ such that $\Phi(\tau) = \sigma^{-1}\tau\sigma$ for every $\tau \in S_3$.

17. Let (1 2) and (1 2 3 $\cdots$ n) be in S_n. Show that any subgroup of S_n that contains both of these must be all of S_n (so these two permutations generate S_n).

18. If τ_1 and τ_2 are two transpositions, show that $\tau_1\tau_2$ can be expressed as the product of 3-cycles (not necessarily disjoint).

19. Prove that if τ_1, τ_2, and τ_3 are transpositions, then $\tau_1\tau_2\tau_3 \neq e$, the identity element of S_n.

20. If τ_1, τ_2 are distinct transpositions, show that $\tau_1\tau_2$ is of order 2 or 3.

21. If σ, τ are two permutations having no letter in common and $\sigma\tau = e$, prove that $\sigma = \tau = e$.

22. Find an algorithm for finding $\sigma\tau\sigma^{-1}$ for any permutations σ, τ of S_n.

23. Let σ, τ be two permutations such that they both have decompositions into disjoint cycles of cycles of lengths $m_1, m_2, \ldots, m_k$. (We say that they have similar decompositions into disjoint cycles.) Prove that for some permutation ρ, $\tau = \rho\sigma\rho^{-1}$.

24. Find the conjugacy class in S_n of $(1 \quad 2 \quad \cdots \quad n)$. What is the order of the centralizer of $(1 \quad 2 \quad \cdots \quad n)$ in S_n?

25. Do Problem 24 for $\sigma = (1 \quad 2)(3 \quad 4)$.

3. ODD AND EVEN PERMUTATIONS

We noticed in Section 2 that although every permutation is the product of transpositions, this decomposition is not unique. We did comment, however, that certain aspects of this kind of decomposition are unique. We go into this now.

Let's consider the special case of S_3, for here we can see everything explicitly. Let $f(x_1, x_2, x_3) = (x_1 - x_2)(x_1 - x_3)(x_2 - x_3)$ be an expression in the three variables x_1, x_2, x_3. We let S_3 act on $f(x) = f(x_1, x_2, x_3)$ as follows. If $\sigma \in S_3$, then

$$\sigma^*(f(x)) = (x_{\sigma(1)} - x_{\sigma(2)})(x_{\sigma(1)} - x_{\sigma(3)})(x_{\sigma(2)} - x_{\sigma(3)}).$$

We consider what σ^* does to $f(x)$ for a few of the σ's in S_3.

Consider $\sigma = (1 \quad 2)$; then $\sigma(1) = 2$, $\sigma(2) = 1$, and $\sigma(3) = 3$, so that

$$\sigma^*(f(x)) = (x_{\sigma(1)} - x_{\sigma(2)})(x_{\sigma(1)} - x_{\sigma(3)})(x_{\sigma(2)} - x_{\sigma(3)})$$

$$= (x_2 - x_1)(x_2 - x_3)(x_1 - x_3)$$

$$= -(x_1 - x_2)(x_1 - x_3)(x_2 - x_3)$$

$$= -f(x).$$

So σ^* *coming from* $\sigma = (1 \quad 2)$ *changes the sign of* $f(x)$. Let's look at the action of another element, $\tau = (1 \quad 2 \quad 3)$, of S_3 on $f(x)$. Then

$$\tau^*(f(x)) = (x_{\tau(1)} - x_{\tau(2)})(x_{\tau(1)} - x_{\tau(3)})(x_{\tau(2)} - x_{\tau(3)})$$

$$= (x_2 - x_3)(x_2 - x_1)(x_3 - x_1)$$

$$= (x_1 - x_2)(x_1 - x_3)(x_2 - x_3)$$

$$= f(x),$$

so τ^* coming from $\tau = (1 \quad 2 \quad 3)$ *leaves* $f(x)$ *unchanged.* What about the other permutations in S_3; how do they affect $f(x)$? Of course, the identity element e induces a map e^* on $f(x)$ which does not change $f(x)$ at all. What does τ^2, for τ above, do to $f(x)$? Since $\tau^* f(x) = f(x)$, we immediately see that

$$(\tau^2)^*(f(x)) = (x_{\tau^2(1)} - x_{\tau^2(2)})(x_{\tau^2(1)} - x_{\tau^2(3)})(x_{\tau^2(2)} - x_{\tau^2(3)})$$

$$= f(x). \quad \text{(Prove!)}$$

Now consider $\sigma\tau = (1 \quad 2)(1 \quad 2 \quad 3) = (2 \quad 3)$; since τ leaves $f(x)$ alone and σ changes the sign of $f(x)$, $\sigma\tau$ must change the sign of $f(x)$. Similarly, $(1 \quad 3)$ changes the sign of $f(x)$. We have accounted for the action of every element of S_3 on $f(x)$.

Suppose that $\rho \in S_3$ is a product $\rho = \tau_1\tau_2 \cdots \tau_k$ of transpositions $\tau_1, \ldots, \tau_k$; then ρ acting on $f(x)$ will change the sign of $f(x)$ k times, since each τ_i changes the sign of $f(x)$. So $\rho^*(f(x)) = (-1)^k f(x)$. If $\rho = \sigma_1\sigma_2 \cdots \sigma_t$, where $\sigma_1, \ldots, \sigma_t$ are transpositions, by the same reasoning, $\rho^*(f(x)) = (-1)^t f(x)$. Therefore, $(-1)^k f(x) = (-1)^t f(x)$, whence $(-1)^t = (-1)^k$. This tells us that t and k have the *same parity*; that is, if t is odd, then k must be odd, and if t is even, then k must be even.

This suggests that although the decomposition of a given permutation σ as a product of transposition is not unique, *the parity of the number of transpositions in such a decomposition of σ might be unique.*

We strive for this goal now, suggesting to readers that they carry out the argument that we do for arbitrary n for the special case $n = 4$.

As we did above, define $f(x) = f(x_1, \ldots, x_n)$ to be

$$f(x) = (x_1 - x_2)(x_1 - x_3) \cdots (x_1 - x_n)(x_2 - x_3)(x_2 - x_4)$$

$$\times \cdots (x_2 - x_n) \cdots (x_{n-1} - x_n)$$

$$= \prod_{i<j}(x_i - x_j),$$

where in this product i takes on all values from 1 to $n - 1$ inclusive, and j all those from 2 to n inclusive. If $\sigma \in S_n$, define σ^* on $f(x)$ by

$$\sigma^*(f(x)) = \prod_{i<j}(x_{\sigma(i)} - x_{\sigma(j)}).$$

If $\sigma, \tau \in S_n$, then

$$(\sigma\tau)^*(f(x)) = \prod_{i<j}(x_{(\sigma\tau)(i)} - x_{(\sigma\tau)(j)}) = \sigma^*\left(\prod_{i<j}(x_{\tau(i)} - x_{\tau(j)})\right)$$

$$= \sigma^*\left(\tau^*\left(\prod_{i<j}(x_i - x_j)\right) = \sigma^*(\tau^*(f(x)) = (\sigma^*\tau^*)(f(x)).$$

So $(\sigma\tau)^* = \sigma^*\tau^*$ when applied to $f(x)$.

What does a transposition τ do to $f(x)$? We claim that $\tau^*(f(x)) = -f(x)$. To prove this, assuming that $\tau = (i \quad j)$ where $i < j$, we count up the number of $(x_u - x_v)$, with $u < v$, which get transformed into an $(x_a - x_b)$ with $a > b$. This happens for $(x_u - x_j)$ if $i < u < j$, for $(x_i - x_v)$ if $i < v < j$, and finally, for $(x_i - x_j)$. Each of these leads to a change of sign on $f(x)$ and since there are $2(j - i - 1) + 1$ such, that is, an odd number of them, we get an odd number of changes of sign on $f(x)$ when acted on by τ^*. Thus $\tau^*(f(x)) = -f(x)$. Therefore, our claim that $\tau^*(f(x)) = -f(x)$ for every transposition τ is substantiated.

If σ is any permutation in S_n and $\sigma = \tau_1\tau_2 \cdots \tau_k$, where $\tau_1, \tau_2, \ldots, \tau_k$ are transpositions, then $\sigma^* = (\tau_1\tau_2 \cdots \tau_k)^* = \tau_1^*\tau_2^* \cdots \tau_k^*$ as acting on $f(x)$, and since each $\tau_i^*(f(x)) = -f(x)$, we see that $\sigma^*(f(x)) = (-1)^k f(x)$. Similarly, if $\sigma = \zeta_1\zeta_2 \cdots \zeta_t$, where $\zeta_1, \zeta_2, \ldots, \zeta_t$ are transpositions, then $\sigma^*(f(x)) = (-1)^t f(x)$. Comparing these two evaluations of $\sigma^*(f(x))$, we conclude that $(-1)^k = (-1)^t$. So these two decompositions of σ as the product of transpositions are of the same parity. *Thus any permutation is either the product of an odd number of transpositions or the product of an even number of transpositions, and no product of an even number of transpositions can equal a product of an odd number of transpositions.*

This suggests the following

Definition. The permutation $\sigma \in S_n$ is an *odd permutation* if σ is the product of an odd number of transpositions, and is an *even permutation* if σ is the product of an even number of transpositions.

What we have proved above is

Theorem 3.3.1. A permutation in S_n is either an odd or an even permutation, *but cannot be both.*

With Theorem 3.3.1 behind us we can deduce a number of its consequences.

Let A_n be the set of all even permutations; if $\sigma, \tau \in A_n$, then we immediately have that $\sigma\tau \in A_n$. Since A_n is thus a finite closed subset of the (finite) group S_n, A_n is a subgroup of S_n, by Lemma 2.3.2. A_n is called the *alternating group of degree n.*

We can show that A_n is a subgroup of S_n in another way. We already saw that A_n is closed under the product of S_n, so to know that A_n is a subgroup of S_n we merely need show that $\sigma \in S_n$ implies that $\sigma^{-1} \in S_n$. For any permutation σ we claim that σ and σ^{-1} are of the same parity. Why? Well, if $\sigma = \tau_1\tau_2 \cdots \tau_k$, where the τ_i are transpositions, then

$$\sigma^{-1} = \left(\tau_1\tau_2 \cdots \tau_k\right)^{-1} = \tau_k^{-1}\tau_{k-1}^{-1} \cdots \tau_2^{-1}\tau_1^{-1} = \tau_k\tau_{k-1} \cdots \tau_2\tau_1,$$

since $\tau_i^{-1} = \tau_i$; therefore, we see that the parity of σ and σ^{-1} is $(-1)^k$, so they are of equal parity. This certainly shows that $\sigma \in A_n$ forces $\sigma^{-1} \in A_n$, whence A_n is a subgroup of S_n.

But it shows a little more, namely that A_n is a *normal subgroup* of S_n. For suppose that $\sigma \in A_n$ and $\rho \in S_n$. What is the parity of $\rho^{-1}\sigma\rho$? By the above, ρ and ρ^{-1} are of the same parity and σ is an even permutation so $\rho^{-1}\sigma\rho$ is an even permutation, hence is in A_n. Thus A_n is a normal subgroup of S_n.

We summarize what we have done in

Theorem 3.3.2. A_n, the alternating group of degree n, is a normal subgroup of S_n.

We look at this in yet another way. From the very definitions involved we have the following simple rules for the product of permutations:

1. The product of two even permutations is even.
2. The product of two odd permutations is even.
3. The product of an even permutation by an odd one (or of an odd one by an even one) is odd.

If σ is an even permutation, let $\theta(\sigma) = 1$, and if σ is an odd permutation, let $\theta(\sigma) = -1$. The foregoing rules about products translate into $\theta(\sigma\tau) = \theta(\sigma)\theta(\tau)$, so θ is a *homomorphism* of S_n onto the group $E = \{1, -1\}$ of order 2 under multiplication. What is the kernel, N, of θ? By the very definition of A_n we see that $N = A_n$. So by the First Homomorphism Theorem, $E \approx S_n/A_n$. Thus $2 = |E| = |S_n/A_n| = |S_n|/|A_n|$, if $n > 1$. This gives us that $|A_n| = \frac{1}{2}|S_n| = \frac{1}{2}n!$.
Therefore,

Theorem 3.3.3.   For $n > 1$, A_n is a normal subgroup of S_n of order $\frac{1}{2}n!$.

Corollary.   In S_n, for $n > 1$, there are $\frac{1}{2}n!$ even permutations and $\frac{1}{2}n!$ odd permutations.

A final few words about the proof of Theorem 3.3.1 before we close this section. Many different proofs of Theorem 3.3.1 are known. Quite frankly, we do not particularly like any of them. Some involve what might be called a "collection process," where one tries to show that e cannot be written as the product of an odd number of transpositions by assuming that it is such a shortest product, and by the appropriate finagling with this product, shortening it to get a contradiction. Other proofs use other devices. The proof we gave exploits the gimmick of the function $f(x)$, which, in some sense, is extraneous to the whole affair. However, the proof given is probably the most transparent of them all, which is why we used it.

Finally, the group A_n, for $n \geq 5$, is an extremely interesting group. We shall show in Chapter 6 that the only normal subgroups of A_n, for

$n \geq 5$, are (e) and A_n itself. A nonabelian group with this property is called a *simple* group (not to be confused with an *easy* group). So the A_n for $n \geq 5$ provide us with an infinite family of simple groups. There are other infinite families of finite simple groups. In the last 20 years or so the heroic efforts of a group of algebraists have determined all finite simple groups. The determination of these simple groups runs about 10,000 printed pages. Interestingly enough, any finite simple group must have *even* order.

Problems

Easier Problems

1. Find the parity of each permutation.

    (a) $\begin{pmatrix} 1 & 2 & 3 & 4 & 5 & 6 & 7 & 8 & 9 \\ 2 & 4 & 5 & 1 & 3 & 7 & 8 & 9 & 6 \end{pmatrix}$.

    (b) $(1\ 2\ 3\ 4\ 5\ 6)(7\ 8\ 9)$.

    (c) $(1\ 2\ 3\ 4\ 5\ 6)(1\ 2\ 3\ 4\ 5\ 7)$.

    (d) $(1\ 2)(1\ 2\ 3)(4\ 5)(5\ 6\ 8)(1\ 7\ 9)$.

2. If σ is a k-cycle, show that σ is an odd permutation if k is even, and is an even permutation if k is odd.

3. Prove that σ and $\tau^{-1}\sigma\tau$, for any $\sigma, \tau \in S_n$, are of the same parity.

4. If $m < n$, we can consider $S_m \subset S_n$ by viewing $\sigma \in S_m$ as acting on $1, 2, \ldots, m, \ldots, n$ as it did on $1, 2, \ldots, n$ and σ leaves $j > m$ fixed. Prove that the parity of a permutation in S_m when viewed this way as an element of S_n does not change.

5. Suppose you are told that the permutation

$$\begin{pmatrix} 1 & 2 & 3 & 4 & 5 & 6 & 7 & 8 & 9 \\ 3 & 1 & 2 & & & 7 & 8 & 9 & 6 \end{pmatrix}$$

in S_9, where the images of 5 and 4 have been lost, is an even permutation. What must the images of 5 and 4 be?

Middle-Level Problems

6. If $n \geq 3$, show that every element in A_n is a product of 3-cycles.

7. Show that every element in A_n is a product of n-cycles.

8. Find a normal subgroup in A_4 of order 4.

Harder Problems (In fact, very hard)

9. If $n \geq 5$ and $(e) \neq N \subset A_n$ is a normal subgroup of A_n, show that N must contain a 3-cycle.

10. Using the result of Problem 9, show that if $n \geq 5$, the only normal subgroups of A_n are (e) and A_n itself. (Thus the groups A_n for $n \geq 5$ give us an infinite family of simple groups.)

CHAPTER **4**

Ring Theory

1. DEFINITIONS AND EXAMPLES

So far in our study of abstract algebra, we have been introduced to one kind of abstract system, which plays a central role in the algebra of today. That was the notion of a group. Because a group is an algebraic system with only one operation, and because a group need' not satisfy the rule $ab = ba$, it ran somewhat counter to our prior experience in algebra. We were used to systems where you could both add and multiply elements and where the elements did satisfy the commutative law of multiplication $ab = ba$. Furthermore, these systems of our acquaintance usually came from sets of numbers—integers, rational, real, and for some, complex.

The next algebraic object we shall consider is a *ring*. In many ways this system will be more reminiscent of what we had previously known than were groups. For one thing rings will be endowed with addition and multiplication, and these will be subjected to many of the familiar rules we all know from arithmetic. On the other hand, rings need not come from our usual number systems, and, in fact, usually have little to do with these familiar ones. Although many of the formal rules of arithmetic hold, many strange—or what may seem as strange—phenomena do take place. As we proceed and see examples of rings, we shall see some of these things occur.

With this preamble over we are ready to begin. Naturally enough, the first thing we should do is to define that which we'll be talking about.

> **Definition.** A nonempty set R is said to be a *ring* if in R there are two operations $+$ and $\cdot$ such that:
>   (a) $a, b \in R$ implies that $a + b \in R$.
>   (b) $a + b = b + a$ for $a, b \in R$.
>   (c) $(a + b) + c = a + (b + c)$ for $a, b, c \in R$.
>   (d) There exists an element $0 \in R$ such that $a + 0 = a$ for every $a \in R$.
>   (e) Given $a \in R$, there exists a $b \in R$ such that $a + b = 0$. (We shall write b as $-a$.)
> Note that so far all we have said is that R is an abelian group under $+$. We now spell out the rules for the multiplication in R.
>   (f) $a, b \in R$ implies that $a \cdot b \in R$.
>   (g) $a \cdot (b \cdot c) = (a \cdot b) \cdot c$ for $a, b, c \in R$.
> This is all that we insist on as far as the multiplication by itself is concerned. But the $+$ and $\cdot$ are not allowed to live in solitary splendor. We interweave them by the two *distributive laws*
>   (h)
>
> $$a \cdot (b + c) = a \cdot b + a \cdot c$$
>
> and
>
> $$(b + c) \cdot a = b \cdot a + c \cdot a,$$
>
> for $a, b, c \in R$.

These axioms for a ring look familiar. They should be, for the concept of ring was introduced as a generalization of what happens in the integers. Because of Axiom (g), the associative law of multiplication, the rings we defined are usually called *associative rings*. Nonassociative rings do exist, and some of these play an important role in mathematics. But they shall not be our concern here. So whenever we use the word "ring" we shall always mean "associative ring."

Although Axioms (a) to (h) are familiar, there are certain things they do not say. We look at some of the familiar rules that are *not insisted* upon for a general ring.

First, we do not postulate the existence of an element $1 \in R$ such that $a \cdot 1 = 1 \cdot a = a$ for every $a \in R$. Many of the examples we shall encounter will have such an element, and in that case we say that *R is a ring with unit*. In all fairness we should point out that many algebraists do demand that a ring have a unit element. We do insist that $1 \neq 0$; that is, the ring consisting of 0 alone is not a ring with unit.

Second, in our previous experience with things of this sort, whenever $a \cdot b = 0$ we concluded that $a = 0$ or $b = 0$. This need not be true, in general, in a ring. When it does hold, the ring is kind of nice and is given a special name: it is called a *domain*.

Third, nothing is said in the axioms for a ring that will imply the *commutative law* of multiplication $a \cdot b = b \cdot a$. There are noncommutative rings where this law does not hold; we shall see some soon. Our main concern in this chapter will be with *commutative rings*, but for many of the early results the commutativity of the ring studied will not be assumed.

As we mentioned above, some things make certain rings nicer than others, and so become worthy of having a special name. We quickly give a list of definitions for some of these nicer rings.

> **Definition.** A commutative ring R is an *integral domain* if $a \cdot b = 0$ in R implies that $a = 0$ or $b = 0$.

It should be pointed out that some algebra books insist that an integral domain contain a unit element. In reading another book, the reader should check if this is the case there. The integers, Z, give us an obvious example of an integral domain. We shall see other, somewhat less obvious, ones.

> **Definition.** A ring R with unit, is said to a *division ring* if for every $a \neq 0$ in R there is an element $b \in R$ (usually written as a^{-1}) such that $a \cdot a^{-1} = a^{-1} \cdot a = 1$.

The reason for calling such a ring a division ring is quite clear, for we can divide (at least keeping left and right sides in mind). Although noncommutative division rings exist with fair frequency and do play an important role in noncommutative algebra, they are fairly complicated and we shall give only one example of these. This division ring is the great classic one introduced by Hamilton in 1843 and is known as the ring of *quaternions*. (See Example 13 below.)

Finally, we come to perhaps the nicest example of a class of rings, the *field*.

> **Definition.**   A ring R is said to be a *field* if R is a *commutative division ring*.

In other words, a field is a commutative ring in which we can divide freely by nonzero elements. Otherwise put, R is a field if the nonzero elements of R form an abelian group under $\cdot$, the product in R.

For fields we do have some ready examples: the rational numbers, the real numbers, the complex numbers. But we shall see many more, perhaps less familiar, examples. Chapter 5 will be devoted to the study of fields.

We spend the rest of the time in this section looking at some examples of rings. *We shall drop the* $\cdot$ *for the product and shall write a* $\cdot$ *b simply as ab.*

EXAMPLES

1. It is obvious which ring we should pick as our first example, namely $\mathbb{Z}$, the ring of integers under the usual addition and multiplication of integers. Naturally enough, $\mathbb{Z}$ is an example of an integral domain.

2. The second example is equally obvious as a choice. Let $\mathbb{Q}$ be the set of all rational numbers. As we all know, $\mathbb{Q}$ satisfies all the rules needed for a field, so $\mathbb{Q}$ is a field.

3. The real numbers, $\mathbb{R}$, also give us an example of a field.

4. The complex numbers, $\mathbb{C}$, form a field.

Note that $\mathbb{Q} \subset \mathbb{R} \subset \mathbb{C}$; we describe this by saying that $\mathbb{Q}$ is a *subfield* of $\mathbb{R}$ (and of $\mathbb{C}$) and $\mathbb{R}$ is a subfield of $\mathbb{C}$.

5. Let $R = \mathbb{Z}_6$, the integers mod 6, with the addition and the multiplication defined by $[a] + [b] = [a + b]$ and $[a][b] = [ab]$.

Note that $[0]$ is the 0 required by our axioms for a ring, and $[1]$ is the unit element of R. Note, however, that $\mathbb{Z}_6$ is *not an integral domain*, for $[2][3] = [6] = [0]$, yet $[2] \neq [0]$ and $[3] \neq [0]$. R is a commutative ring with unit.

This example suggests the

Definition. An element $a \neq 0$ in a ring R is a *zero-divisor* in R if $ab = 0$ for some $b \neq 0$ in R.

We should really call what we defined a left zero-divisor; however, since we shall mainly talk about commutative rings, we shall not need any left–right distinction for zero-divisors.

Note that both $[2]$ and $[3]$ in $\mathbb{Z}_6$ are zero-divisors. An integral domain is, of course, a commutative ring without zero-divisors.

6. Let $R = \mathbb{Z}_5$, the ring of integers mod 5. R is, of course, a commutative ring with unit. But it is more; in fact, it is a field. Its nonzero elements are $[1], [2], [3], [4]$ and we note that $[2][3] = [6] = [1]$, and $[1]$ and $[4]$ are their own inverses. So every nonzero element in $\mathbb{Z}_5$ has an inverse in $\mathbb{Z}_5$.

We generalize (6) for any prime p.

7. Let $\mathbb{Z}_p$ be the integers mod p, where p is a prime. Again $\mathbb{Z}_p$ is clearly a commutative ring with 1. We claim that $\mathbb{Z}_p$ is a field. To see this, note that if $[a] \neq [0]$, then $p \nmid a$. Therefore, by Fermat's Theorem (Corollary to Theorem 2.4.8), $a^{p-1} \equiv 1(p)$. For the classes $[\cdot]$ this says that $[a^{p-1}] = [1]$. But $[a^{p-1}] = [a]^{p-1}$, so $[a]^{p-1} = [1]$; therefore, $[a]^{p-2}$ is the required inverse for $[a]$ in $\mathbb{Z}_p$, hence $\mathbb{Z}_p$ is a field.

Because $\mathbb{Z}_p$ has only a finite number of elements, it is called a *finite field*. Later we shall construct finite fields different from the $\mathbb{Z}_p$'s.

8. Let $\mathbf{Q}$ be the rational numbers; if $a \in \mathbf{Q}$, we can write $a = m/n$, where m are n are relatively prime integers. Call this the *reduced form* for a. Let R be the set of all $a \in \mathbf{Q}$ in whose reduced form the denominator is odd. Under the usual addition and multiplication in $\mathbf{Q}$ the set R forms a ring. It is an integral domain with unit but is not a field, for $\frac{1}{2}$, the needed inverse of 2, is not in R. Exactly which elements in R do have their inverses in R?

9. Let R be the set of all $a \in \mathbf{Q}$ in whose reduced form the denominator is not divisible by a fixed prime p. As in (8), R is a ring under the usual addition and multiplication in $\mathbf{Q}$, is an integral domain but is not a field. What elements of R have their inverses in R?

Definition.   Both Examples 8 and 9 are subrings of $\mathbf{Q}$ in the following sense. R is a ring, then a *subring* of R is a subset S of R which is a ring if the operations ab and $a + b$ ($a, b \in S$) are just the operations of R applied to the elements $a, b \in S$.

For S to be a subring, it is necessary and sufficient that S be nonempty and that $ab, a \pm b \in S$ for all $a, b \in S$. (Prove!)

We give one further commutative example. This one comes from the calculus.

10. Let R be the set of all real-valued continuous functions on the closed unit interval $[0, 1]$. For $f, g \in R$ and $x \in [0, 1]$ define $(f + g)(x) = f(x) + g(x)$, and $(f \cdot g)(x) = f(x)g(x)$. From the results in the calculus, $f + g$ and $f \cdot g$ are again continuous functions on $[0, 1]$. With these operations R is a commutative ring. It is *not* an integral domain. For instance, if $f(x) = -x + \frac{1}{2}$ for $0 \leq x \leq \frac{1}{2}$ and $f(x) = 0$ for $\frac{1}{2} < x \leq 1$, and if $g(x) = 0$ for $0 \leq x \leq \frac{1}{2}$ and $g(x) = 2x - 1$ for $\frac{1}{2} < x \leq 1$, then $f, g \in R$ and, as is easy to verify, $f \cdot g = 0$. It does have a unit element, namely the function e defined by $e(x) = 1$ for all $x \in [0, 1]$. What elements of R have their inverses in R?

We should now like to see some noncommutative examples. These are not so easy to come by, although noncommutative rings exist in abundance because we are not assuming any knowledge of linear algebra on the reader's part. The easiest and most natural first source of such examples is the set of matrices over a field. So, in our first noncommutative example, we shall really create the 2×2 matrices with real entries.

11. Let F be the field of real numbers and let R be the set of all formal square arrays

$$\begin{pmatrix} a & b \\ c & c \end{pmatrix}$$

where a, b, c, d are any real numbers. For such square arrays we define addition in a natural way by defining

$$\begin{pmatrix} a_1 & b_1 \\ c_1 & d_1 \end{pmatrix} + \begin{pmatrix} a_2 & b_2 \\ c_2 & d_2 \end{pmatrix} = \begin{pmatrix} a_1 + a_2 & b_1 + b_2 \\ c_1 + c_2 & d_1 + d_2 \end{pmatrix}.$$

It is easy to see that R forms an abelian group under this $+$ with $\begin{pmatrix} 0 & 0 \\ 0 & 0 \end{pmatrix}$ acting as the zero element and $\begin{pmatrix} -a & -b \\ -c & -d \end{pmatrix}$ the negative of $\begin{pmatrix} a & b \\ c & d \end{pmatrix}$. To make of R a ring, we need a multiplication. We define one in what may seem a highly unnatural way via

$$\begin{pmatrix} a & b \\ c & d \end{pmatrix} \begin{pmatrix} r & s \\ t & u \end{pmatrix} = \begin{pmatrix} ar + bt & as + bu \\ cr + dt & cs + du \end{pmatrix}.$$

It may be a little laborious, but one can check that with these operations R is a noncommutative ring with $\begin{pmatrix} 1 & 0 \\ 0 & 1 \end{pmatrix}$ acting as its multiplicative unit element. Note that

$$\begin{pmatrix} 1 & 0 \\ 0 & 0 \end{pmatrix} \begin{pmatrix} 0 & 0 \\ 1 & 0 \end{pmatrix} = \begin{pmatrix} 0 & 0 \\ 0 & 0 \end{pmatrix}$$

while

$$\begin{pmatrix} 0 & 0 \\ 1 & 0 \end{pmatrix} \begin{pmatrix} 1 & 0 \\ 0 & 0 \end{pmatrix} = \begin{pmatrix} 0 & 0 \\ 1 & 0 \end{pmatrix},$$

so

$$\begin{pmatrix} 1 & 0 \\ 0 & 0 \end{pmatrix} \begin{pmatrix} 0 & 0 \\ 1 & 0 \end{pmatrix} \neq \begin{pmatrix} 0 & 0 \\ 1 & 0 \end{pmatrix} \begin{pmatrix} 1 & 0 \\ 0 & 0 \end{pmatrix}.$$

Note that $\begin{pmatrix} 1 & 0 \\ 0 & 0 \end{pmatrix}$ and $\begin{pmatrix} 0 & 0 \\ 1 & 0 \end{pmatrix}$ are zero-divisors; in fact,

$$\begin{pmatrix} 0 & 0 \\ 1 & 0 \end{pmatrix}^2 = \begin{pmatrix} 0 & 0 \\ 0 & 0 \end{pmatrix} \quad \text{so} \quad \begin{pmatrix} 0 & 0 \\ 1 & 0 \end{pmatrix}$$

is a nonzero element whose square is the 0 element of R. R is known as the *ring of all* 2×2 *matrices over* F, *the real field*.

For those unfamiliar with these matrices, and who see no sense in the product defined for them, let's look at how we do compute the product. To get the top left entry in the product AB, we "multiply" the first row of A by the first column of B, where $A, B \in R$. For the top right entry, it is the first row of A versus the second column of B. The bottom left entry comes from the second row of A versus the first column of B, and finally, the bottom right entry is the second column of A versus the second column of B.

We illustrate with an example: Let

$$A = \begin{pmatrix} 1 & \frac{1}{2} \\ -3 & 2 \end{pmatrix} \quad \text{and} \quad B = \begin{pmatrix} \frac{1}{3} & \frac{2}{5} \\ \pi & -\pi \end{pmatrix}.$$

Then the first row of A is $1, \frac{1}{2}$ and the first column of B is $\frac{1}{3}, \pi$; we "multiply" these via $1 \cdot \frac{1}{3} + \frac{1}{2} \cdot \pi = \pi/2 + \frac{1}{3}$, and so on. So we see that

$$AB = \begin{pmatrix} \frac{1}{3} + \pi/2 & \frac{2}{5} - \pi/2 \\ -1 + 2\pi & -\frac{6}{5} - 2\pi \end{pmatrix}.$$

In the problems we shall have many matrix multiplications, so that the reader can acquire some familiarity with this strange but important example.

12. Let R be any ring and let

$$S = \left\{ \begin{pmatrix} a & b \\ c & d \end{pmatrix} \middle| a, b, c, d \in R \right\}$$

with $+$ and $\cdot$ as defined in Example 10. One can verify that S is a ring, also, under these operations. It is called the *ring of* 2×2 *matrices over* R.

Our final example is one of the great classical examples, the *real quaternions*, introduced by Hamilton (as a noncommutative parallel to the complex numbers).

13. *The quaternions.* Let F be the field of real numbers and consider the set of all formal symbols $\alpha_0 + \alpha_1 i + \alpha_2 j + \alpha_3 k$, where $\alpha_0, \alpha_1, \alpha_2, \alpha_3 \in F$. Equality and addition of these symbols are easy, via the obvious route

$$\alpha_0 + \alpha_1 i + \alpha_2 j + \alpha_3 k = \beta_0 + \beta_1 i + \beta_2 j + \beta_3 k$$

if and only if $\alpha_0 = \beta_0$, $\alpha_1 = \beta_1$, $\alpha_2 = \beta_2$ and $\alpha_3 = \beta_3$, and

$$(\alpha_0 + \alpha_1 i + \alpha_2 j + \alpha_3 k) + (\beta_0 + \beta_1 i + \beta_2 j + \beta_3 k)$$

$$= (\alpha_0 + \beta_0) + (\alpha_1 + \beta_1) i + (\alpha_2 + \beta_2) j + (\alpha_3 + \beta_3) k.$$

We now come to the tricky part, the multiplication. When Hamilton discovered it on October 6, 1843, he cut the basic rules of this product out with his penknife on Brougham Bridge in Dublin. The product is based on $i^2 = j^2 = k^2 = -1$, $ij = k$, $jk = i$, $ki = j$ and $ji = -k$, $kj = -i$, $ik = -j$. If we go around the circle clockwise

the product of any two successive ones is the next one, and going around counterclockwise we get the negatives.

We can write out the product now of any two quaternions, according to the rules above, *declaring by definition* that

$$(\alpha_0 + \alpha_1 i + \alpha_2 j + \alpha_3 k)(\beta_0 + \beta_1 i + \beta_2 j + \beta_3 k)$$

$$= \gamma_0 + \gamma_1 i + \gamma_2 j + \gamma_3 k,$$

where

(I)

$$\gamma_0 = \alpha_0 \beta_0 - \alpha_1 \beta_1 - \alpha_2 \beta_2 - \alpha_3 \beta_3$$

$$\gamma_1 = \alpha_0 \beta_1 + \alpha_1 \beta_0 + \alpha_2 \beta_3 - \alpha_3 \beta_2$$

$$\gamma_2 = \alpha_c \beta_2 - \alpha_1 \beta_3 + \alpha_2 \beta_0 + \alpha_3 \beta_1$$

$$\gamma_3 = \alpha_0 \beta_3 + \alpha_1 \beta_2 - \alpha_2 \beta_1 + \alpha_3 \beta_0$$

It looks horrendous, doesn't it? But it's not as bad as all that. We are multiplying out formally using the distributive laws and using the product rules for the i, j, k above.

If some α_i is 0 in $x = \alpha_0 + \alpha_1 i + \alpha_2 j + \alpha_3 k$, we shall omit it in expressing x; thus $0 + 0i + 0j + 0k$ will be written simply as 0, $1 + 0i + 0j + 0k$ as 1, $0 + 3i + 4j + 0k$ as $3i + 4j$, and so on.

A calculation reveals that

(II)

$$(\alpha_0 + \alpha_1 i + \alpha_2 j + \alpha_3 k)(\alpha_0 - \alpha_1 i - \alpha_2 j - \alpha_3 k)$$

$$= \alpha_0^2 + \alpha_1^2 + \alpha_2^2 + \alpha_3^2$$

This has a very important consequence; for suppose that $x = \alpha_0 + \alpha_1 i + \alpha_2 j + \alpha_3 k \neq 0$ (so some $\alpha_i \neq 0$). Then, since the α's are real, $\beta = \alpha_0^2 + \alpha_1^2 + \alpha_2^2 + \alpha_3^2 \neq 0$. Then from (II) we easily get

$$(\alpha_0 + \alpha_1 i + \alpha_2 j + \alpha_3 k)\left(\frac{\alpha_0}{\beta} - \frac{\alpha_1}{\beta} i - \frac{\alpha_2}{\beta} j - \frac{\alpha_3}{\beta} k\right) = 1.$$

So, if $x \neq 0$, then x has an inverse in the quaternions. *Thus the quaternions form a noncommutative division ring.*

Although, as we mentioned earlier, there is no lack of noncommutative division rings, the quaternions above (or some piece of them) are often the only noncommutative division rings that even many professional mathematicians have ever seen.

We shall have many problems—some easy and some quite a bit harder—about the two examples: the 2×2 matrices and the quaternions. This way the reader will be able to acquire some skill with playing with noncommutative rings.

One final comment in this section: If $\gamma_0, \gamma_1, \gamma_2, \gamma_3$ are as in (I), then

(III)

$$(\alpha_0^2 + \alpha_1^2 + \alpha_2^2 + \alpha_3^2)(\beta_0^2 + \beta_1^2 + \beta_2^2 + \beta_3^2)$$

$$= \gamma_0^2 + \gamma_1^2 + \gamma_2^2 + \gamma_3^2.$$

This is known as *Lagrange's Identity*; it expresses the product of two sums of four squares again as a sum of four squares. Its verification will be one of the exercises.

Problems

Easier Problems

*1. Find all the elements in $\mathbb{Z}_{24}$ that are *invertible* (i.e., have multiplicative inverse) in $\mathbb{Z}_{24}$.

2. Show that any field is an integral domain.

3. Show that $\mathbb{Z}_n$ is a field if and only if n is a prime.

4. Verify that Example 8 is a ring. Find all its invertible elements.

5. Do Problem 4 for Example 9.

6. In Example 10, the 2×2 matrices over the reals, check the associative law of multiplication.

7. Work out the following:

(a) $\begin{pmatrix} 1 & 2 \\ 4 & -7 \end{pmatrix}\begin{pmatrix} \frac{1}{5} & \frac{2}{3} \\ 0 & 1 \end{pmatrix}$.

(b) $\begin{pmatrix} 1 & 1 \\ 1 & 1 \end{pmatrix}^2$.

(c) $\begin{pmatrix} \frac{1}{2} & \frac{1}{2} \\ 0 & 0 \end{pmatrix}^3$.

(d) $\begin{pmatrix} a & b \\ c & d \end{pmatrix}\begin{pmatrix} 1 & 0 \\ 0 & 0 \end{pmatrix} - \begin{pmatrix} 1 & 0 \\ 0 & 0 \end{pmatrix}\begin{pmatrix} a & b \\ c & d \end{pmatrix}$.

8. Find all matrices $\begin{pmatrix} a & b \\ c & d \end{pmatrix}$ such that $\begin{pmatrix} a & b \\ c & d \end{pmatrix}\begin{pmatrix} 1 & 0 \\ 0 & 0 \end{pmatrix} = \begin{pmatrix} 1 & 0 \\ 0 & 0 \end{pmatrix}\begin{pmatrix} a & b \\ c & d \end{pmatrix}$.

9. Find all 2×2 matrices $\begin{pmatrix} a & b \\ c & d \end{pmatrix}$ that commute with all 2×2 matrices.

10. Let R be any ring with unit, S the ring of 2×2 matrices over R. (See Example 11.)

(a) Check the associative law of multiplication in S. (**Remember:** R need not be commutative.)

(b) Show that $\left\{ \begin{pmatrix} a & b \\ 0 & c \end{pmatrix} \middle| a, b, c, \in R \right\}$ is a subring of S.

(c) Show that $\begin{pmatrix} a & b \\ 0 & c \end{pmatrix}$ has an inverse in S if and only if a and c have inverses in R. In that case write down $\begin{pmatrix} a & b \\ 0 & c \end{pmatrix}^{-1}$ explicitly.

11. Let $F: \mathbf{C} \to \mathbf{C}$ be defined by $F(a + bi) = a - bi$. Show that:
   (a) $F(xy) = F(x)F(y)$ for $x, y \in \mathbf{C}$.
   (b) $F(x\bar{x}) = |x|^2$.
   (c) Using Parts (a) and (b), show that
   $$(a^2 + b^2)(c^2 + d^2) = (ac - bd)^2 + (ad + bc)^2$$
   [**Note:** $F(x)$ is merely $\bar{x}$.]

12. Verify the identity in Part (b) of Problem 11 directly.

13. Find the following products of quaternions.
   (a) $(i + j)(i - j)$.
   (b) $(1 - i + 2j - 2k)(1 + 2i - 4j + 6k)$.
   (c) $(2i - 3j + 4k)^2$.
   (d) $i(\alpha_0 + \alpha_1 i + \alpha_2 j + \alpha_3 k) - (\alpha_0 + \alpha_1 i + \alpha_2 j + \alpha_3 k)i$.

14. Show that the only quaternions commuting with i are of the form $\alpha + \beta i$.

15. Find the quaternions that commute with both i and j.

16. Verify that
   $$(\alpha_0 + \alpha_1 i + \alpha_2 j + \alpha_3 k)(\alpha_0 - \alpha_1 i - \alpha_2 j - \alpha_3 k)$$
   $$= \alpha_0^2 + \alpha_1^2 + \alpha_2^2 + \alpha_3^2.$$

17. Verify Lagrange's Identity by a direct calculation.

Middle-Level Problems

18. In the quaternions, define
   $$|\alpha_0 + \alpha_1 i + \alpha_2 j + \alpha_3 k| = \sqrt{\alpha_0^2 + \alpha_1^2 + \alpha_2^2 + \alpha_3^2}.$$

Show that $|xy| = |x| |y|$ for any two quaternions x and y.

19. Show that there is an *infinite* number of solutions to $x^2 = -1$ in the quaternions.

20. In the quaternions, consider the following set G having eight elements: $G = \{\pm 1, \pm i, \pm j, \pm k\}$.
(a) Prove that G is a group (under multiplication).
(b) List all subgroups of G.
(c) What is the center of G?
(d) Show that G is a *nonabelian group all of whose subgroups are normal.*

21. Show that a division ring is a domain.

22. Give an example, in the quaternions, of a noncommutative domain that is not a division ring.

23. *Define* the map * in the quaternions by

$$(\alpha_0 + \alpha_1 i + \alpha_2 j + \alpha_3 k)^* = (\alpha_0 - \alpha_1 i - \alpha_2 j - \alpha_3 k).$$

Show that:
(a) $x^{**} = (x^*)^* = x$.
(b) $(x + y)^* = x^* + y^*$.
(c) $xx^* = x^*x$ is real and nonnegative.
(d) $(xy)^* = y^*x^*$.
[Note the reversal of order in Part (d).]

24. Using *, define $|x| = \sqrt{xx^*}$. Show that $|xy| = |x| |y|$ for any two quaternions x and y, by using Parts (c) and (d) of Problem 23.

25. Use the result of Problem 24 to prove Lagrange's Identity.

In Problems 26 to 30, let R be the 2×2 matrices over the reals.

26. If $\begin{pmatrix} a & b \\ c & d \end{pmatrix} \in R$, show that $\begin{pmatrix} a & b \\ c & d \end{pmatrix}$ is invertible in R if and only if $ad - bc \neq 0$. In that case find $\begin{pmatrix} a & b \\ c & d \end{pmatrix}^{-1}$

27. Define $\det \begin{pmatrix} a & b \\ c & d \end{pmatrix} = ad - bc$. For $x, y \in R$ show that $\det(xy) = (\det x)(\det y)$.

28. Show that $\{x \in R \mid \det x \neq 0\}$ forms a group, G, under matrix multiplication and that $N = \{x \in R \mid \det x = 1\}$ is a normal subgroup of G.

29. If $x \in R$ is a zero-divisor, show that $\det x = 0$, and, conversely, if $x \neq 0$ is such that $\det x = 0$, then x is a zero-divisor in R.

30. In R, show that $\left\{ \begin{pmatrix} a & b \\ -b & a \end{pmatrix} \middle| a, b \text{ real} \right\}$ is a field.

Harder Problems

31. Let R be the ring of all 2×2 matrices over $\mathbb{Z}_p$, p a prime. Show that if $\det \begin{pmatrix} a & b \\ c & d \end{pmatrix} = ad - bc \neq 0$, then $\begin{pmatrix} a & b \\ c & d \end{pmatrix}$ is invertible in R.

32. Let R be as in Problem 31. Show that for $x, y \in R$, $\det(xy) = \det(x)\det(y)$.

33. Let G be the set of elements in R of Problem 31 such that $\det(x) \neq 0$.
    (a) Prove that G is a group.
    (b) Find the order of G. (*Quite hard*)
    (c) Find the center of G.
    (d) Find a p-Sylow subgroup of G.

34. Let T be the group of matrices A with entries in the field $\mathbb{Z}_2$ such that $\det A$ is not equal to 0. Prove that T is isomorphic to S_3, the symmetric group of degree 3.

35. Show that the ring R in Example 9 (continuous functions on $[0, 1]$) is *not* an integral domain.

If F is a field, let $H(F)$ be the ring of quaternions over F, that is, the set of all $\alpha_0 + \alpha_1 i + \alpha_2 j + \alpha_3 k$, where $\alpha_0, \alpha_1, \alpha_2, \alpha_3 \in F$ and where equality, addition, and multiplication are defined as for the real quaternions.

36. If $F = \mathbb{C}$, the complex numbers, show that $H(\mathbb{C})$ is *not* a division ring.

37. In $H(\mathbf{C})$, find an element $x \neq 0$ such that $x^2 = 0$.

38. Show that $H(F)$ is a division ring if and only if $\alpha_0^2 + \alpha_1^2 + \alpha_2^2 + \alpha_3^2 = 0$ for $\alpha_1, \alpha_2, \alpha_3, \alpha_4$ in F forces $\alpha_0 = \alpha_1 = \alpha_2 = \alpha_3 = 0$.

39. If Q is the field of rational numbers, show that $H(Q)$ is a division ring.

40. Prove that a finite domain is a division ring.

41. Use Problem 40 to show that $\mathbb{Z}_p$ is a field if p is a prime.

2. SOME SIMPLE RESULTS

Now that we have seen some examples of rings and have had some experience playing around with them, it would seem wise to develop some computational rules. These will allow us to avoid annoying trivialities that could beset a calculation we might be making.

The results we shall prove in this section are not very surprising, not too interesting, and certainly not at all exciting. Neither was learning the alphabet, but it was something we had to do before going on to bigger and better things. The same holds for the results we are about to prove.

Since a ring R is at least an abelian group under $+$, there are certain things we know from our group theory background, for instance, $-(-a) = a$, $-(a + b) = (-a) + (-b)$; if $a + b = a + c$, then $b = c$, and so on.

We begin with

Lemma 4.2.1. Let R be any ring and let $a, b \in R$. Then
   (a) $a0 = 0a = 0$.
   (b) $a(-b) = (-a)b = -(ab)$.
   (c) $(-a)(-b) = ab$.
   (d) If $1 \in R$, then $(-1)a = -a$.

Proof. We do these in turn.
   (a) Since $0 = 0 + 0$, $a0 = a(0 + 0) = a0 + a0$, hence $a0 = 0$. We have used the left distributive law in this proof. The right distributive law gives $0a = 0$.

(b) $ab + a(-b) = a(b + (-b)) = a0 = 0$ from Part (a). Therefore, $a(-b) = -(ab)$. Similarly, $(-a)b = -(ab)$.

(c) By Part (b), $(-a)(-b) = -((-a)b) = -(-(ab)) = ab$, since we are in an abelian group.

(d) If $1 \in R$, then $(-1)a + a = (-1)a + (1)a = (-1 + 1)a = 0a = 0$. So $(-1)a = -a$ by the definition of $-a$. $\square$

Another computational result.

Lemma 4.2.2.   In any ring $R, (a + b)^2 = a^2 + b^2 + ab + ba$ for $a, b \in R$.

Proof.   This is clearly the analog of $(\alpha + \beta)^2 = \alpha^2 + 2\alpha\beta + \beta^2$ in the integers, say, but keeping in mind that R may be noncommutative. So, to it. By the right distributive law $(a + b)^2 = (a + b)(a + b) = (a + b)a + (a + b)b = a^2 + ba + ab + b^2$, exactly what was claimed. $\square$

Can you see the noncommutative version of the binomial theorem? Try it for $(a + b)^3$.

One curiosity follows from the two distributive laws when R has a unit element. The commutative law of addition follows from the rest.

Lemma 4.2.3.   If R is a system with 1 satisfying all the axioms of a ring, except possibly $a + b = b + a$ for $a, b \in R$, then R is a ring.

Proof.   We must show that $a + b = b + a$ for $a, b \in R$. By the right distributive law $(a + b)(1 + 1) = (a + b)1 + (a + b)1 = a + b + a + b$. On the other hand, by the left distributive law $(a + b)(1 + 1) = a(1 + 1) + b(1 + 1) = a + a + b + b$. But then $a + b + a + b = a + a + b + b$; since we are in a group under $+$, we can cancel a on the left and b on the right to obtain $b + a = a + b$, as required. R is therefore a ring. $\square$

We close this brief section with a result that is a little nicer. We say that a ring R is a *Boolean ring* [after the English mathematician George Boole (1815–1864)] if $x^2 = x$ for every $x \in R$.

We prove the cute

Lemma 4.2.4.   A Boolean ring is commutative.

Proof.   Let $x, y \in R$, a Boolean ring. Thus $x^2 = x$, $y^2 = y$, $(x + y)^2 = x + y$. But $(x + y)^2 = x^2 + xy + yx + y^2 = x + xy + yx + y$, by Lemma 4.2.2, so we have $(x + y) = (x + y)^2 = x + xy + yx + y$, from which we have $xy + yx = 0$.

Thus $0 = x(xy + yx) = x^2y + xyx = xy + xyx$, while

$$0 = (xy + yx)x = xyx + yx^2 = xyx + yx.$$

This gives us $xy + xyx = xyx + yx$, and so $xy = yx$. Therefore, R is commutative. □

Problems

1. Let R be a ring; since R is an abelian group under $+$, na has meaning for us for $n \in \mathbb{Z}$, $a, b \in R$. Show that $(na)(mb) = (nm)(ab)$ if n, m are integers and $a, b \in R$.

2. If R is an integral domain and $ab = ac$ for $a \neq 0$, $b, c \in R$, show that $b = c$.

3. If R is a finite integral domain, show that R is a field.

4. If R is a ring and $e \in R$ is such that $e^2 = e$, show that $(xe - exe)^2 = (ex - exe)^2 = 0$ for every $x \in R$.

5. Let R be a ring in which $x^3 = x$ for every $x \in R$. Prove that R is commutative.

6. If $a^2 = 0$ in R, show that $ax + xa$ commutes with a.

7. Let R be a ring in which $x^4 = x$ for every $x \in R$. Prove that R is commutative.

8. If F is a finite field, show that:
   (a) There exists a prime p such that $pa = 0$ for all $a \in F$.
   (b) If F has q elements, then $q = p^n$ for some integer n. (**Hint:** Cauchy's Theorem?)

9. Let p be an odd prime and let $1 + \frac{1}{2} + \cdots + 1/(p-1) = a/b$, where a, b are integers. Show that $p|a$. (**Hint:** As a runs through $\mathbb{Z}_p$, so does a^{-1}.)

10. If p is a prime and $p > 3$, show that if $1 + \frac{1}{2} + \cdots + 1/(p-1) = a/b$, where a, b are integers, then $p^2|a$. (**Hint:** Consider $1/a^2$ as a runs through $\mathbb{Z}_p$.)

3.   IDEALS, HOMOMORPHISMS, AND QUOTIENT RINGS

In studying groups, it turned out that homomorphisms, and their kernels —the normal subgroups—played a central role. There is no reason to expect that the same thing should not be true for rings. As a matter of fact, the analogs, in the setting of rings, of homomorphism and normal subgroup do play a key role.

With the background we have acquired about such things in group theory, the parallel development for rings should be easy and quick. And it will be! Without any further fuss we make the

> **Definition.**   The mapping $\varphi: R \to R'$ of the ring R into the ring R' is a *homomorphism* if
> (a) $\varphi(a + b) = \varphi(a) + \varphi(b)$ and
> (b) $\varphi(ab) = \varphi(a)\varphi(b)$ for all $a, b \in R$.

Since a ring has two operations, it is only natural and just that we demand that both these operations be preserved under what we would call a ring homomorphism. *Furthermore, Property (a) in the Definition tells us that φ is a homomorphism of R viewed merely as an abelian group under $+$ into R'* (also viewed as a group under its addition). So we can call on, and expect, certain results from this fact alone.

Just as we saw in Chapter 2, Section 5 for groups, the image of R under a homomorphism from R to R', is a subring of R', as defined in Chapter 4, Section 1 (Prove!).

Let $\varphi: R \to R'$ be a ring homomorphism and let $\text{Ker } \varphi = \{x \in R | \varphi(x) = 0\}$, the 0 being that of R'. What properties does $\text{Ker } \varphi$ enjoy? Clearly, from group theory $\text{Ker } \varphi$ is an additive subgroup of R. But much more is true. If $k \in \text{Ker } \varphi$ and $r \in R$, then $\varphi(k) = 0$, so

$\varphi(kr) = \varphi(k)\varphi(r) = 0\varphi(r) = 0$, and similarly, $\varphi(rk) = 0$. So Ker φ swallows up multiplication from the left and the right by arbitrary ring elements.

This property of Ker φ is now abstracted to define the important analog in ring theory of the notion of normal subgroup in group theory.

> **Definition.** Let R be a ring. A nonempty subset I of R is called an *ideal* of R if:
> (a) I is an additive subgroup of R.
> (b) Given $r \in R$, $a \in I$, then $ra \in I$ and $ar \in I$.

We shall soon see some examples of homomorphisms and ideals. But first we note that Part (b) in the definition of ideal really has a left and a right part. We could split it and define a set L of R to be a *left ideal* of R if L is an additive subgroup of R and given $r \in R$, $a \in L$, then $ra \in L$. So we require only left swallowing-up for a left ideal. We can similarly define *right ideals*. An ideal as we defined it is simultaneously both a left and a right ideal of R. By all rights we should then call what we called an ideal a *two-sided ideal* of R. Indeed, in working in noncommutative ring theory one uses this name; *here, by "ideal" we shall always mean a two-sided ideal.* Except for some of the problems, we shall not use the notion of one-sided ideals in this chapter.

Before going on, we record what was done above for Ker φ as

> **Lemma 4.3.1.** If $\varphi: R \to R'$ is a homomorphism, then Ker φ is an ideal of R.

We shall soon see that every ideal can be made the kernel of a homomorphism. Shades of what happened for normal subgroups of groups!

Finally, let K be an ideal of R. Since K is an additive subgroup of R, the quotient group R/K exists; it is merely the set of all cosets $a + K$ as a runs over R. But R is not just a group; it is, after all, a ring. Nor is K merely an additive subgroup of R; it is more than that, namely an ideal of R. We should be able to put all this together to make of R/K a ring.

How should we define a product in R/K in a natural way? What do we want to declare $(a + K)(b + K)$ to be? The reasonable thing is to define $(a + K)(b + K) = ab + K$, which we do. As always, the first thing that comes up is to show that this product is well-defined. Is it? We must show that if $a + K = a' + K$ and $b + K = b' + K$, then $(a + K)(b + K) = ab + K = a'b' + K = (a' + K)(b' + K)$. However, if $a + K = a' + K$, then $a - a' \in K$, so $(a - a')b \in K$, since K is an ideal of R (in fact, so far, since K is a right ideal of R). Because $b + K = b' + K$, we have $b - b' \in K$, so $a'(b - b') \in K$, since K is an ideal of R (in fact, since K is a left ideal of R). So both

$$(a - a')b = ab - a'b \qquad \text{and} \qquad a'(b - b') = a'b - a'b'$$

are in K. Thus

$$(ab - a'b) + (a'b - a'b') = ab - a'b' \in K.$$

But this tells us (just from group theory) that $ab + K = a'b' + K$, exactly what we needed to have the product well defined.

So R/K is now endowed with a sum and a product. Furthermore, the mapping $\varphi : R \to R/K$ defined by $\varphi(a) = a + K$ for $a \in R$ is a homomorphism of R onto R/K with kernel K. (Prove!) This tells us right away that R/K is a ring, being the homomorphic image of the ring R.

We summarize all this in

> **Theorem 4.3.2.** Let K be an ideal of R. Then the quotient group R/K as an additive group is a ring under the multiplication $(a + K)(b + K) = ab + K$. Furthermore, the map $\varphi : R \to R/K$ defined by $\varphi(a) = a + K$ for $a \in R$ is a homomorphism of R onto R/K having K as its kernel. So R/K is a homomorphic image of R.

Just from group-theoretic consideration of R as an additive group, we have that if φ is a homomorphism of R into R', then it is 1-1 if and only if $\text{Ker}\,\varphi = (0)$. As in groups we define a homomorphism to be a *monomorphism* if it is 1-1. A monomorphism which is also onto is called an *isomorphisim*. We define R and R' to be *isomorphic* if there is an *isomorphism* of R onto R'.

An *isomorphism* from a ring R onto itself is called an *automorphism* of R. For example, if R is the field $\mathbb{C}$ of complex numbers, then the mapping from R to R sending each element of R to its complex conjugate is an automorphism of $\mathbb{C}$. (Prove!)

One would have to be an awful pessimist not to expect that the homomorphism theorems proved in Sections 5 and 6 of Chapter 2 hold. In fact they do, with the slightly obvious adaptation needed to make the proofs go through. We state the homomorphism theorems without any further ado, leaving the few details needed to complete the proofs to the reader.

Theorem 4.3.3 (First Homomorphism Theorem). Let the mapping $\varphi: R \to R'$ be a homomorphism of R onto R' with kernel K. Then $R' \simeq R/K$; in fact, the mapping $\psi: R/K \to R'$ defined by $\psi(a + K) = \varphi(a)$ defines an isomorphism of R/K onto R'.

We go on to the next homomorphism theorem.

Theorem 4.3.4 (Second Homomorphism Theorem). Let the mapping $\varphi: R \to R'$ be a homomorphism of R onto R' with kernel K. If I' is an ideal of R', let $I = \{a \in R | \varphi(a) \in I'\}$. Then I is an ideal of R, $I \supset K$ and $I/K \simeq I'$. This sets up a 1-1 correspondence between all the ideals of R' and those ideals of R that contain K.

Finally, we come to the last of the homomorphism theorems that we wish to state. There are, of course, others; we leave those to the problems.

Theorem 4.3.5 (Third Homomorphism Theorem). Let the mapping $\varphi: R \to R'$ be a homomorphism of R onto R' with kernel K. If I' is an ideal of R' and $I = \{a \in R | \varphi(a) \in I'\}$, then $R/I \simeq R'/I'$. Equivalently, if K is an ideal of R and $I \supset K$ is an ideal of R, then $R/I \simeq (R/K)/(I/K)$.

We close this section with an inspection of some of the things we have discussed in some examples.

EXAMPLES

1. As usual we use Z, the ring of integers, for our first example. Let $n > 1$ be a fixed integer and let I_n be the set of all multiples of n; then I_n is an ideal of R. If Z_n is the integers mod n, define $\varphi: Z \to Z_n$ by $\varphi(a) = [a]$. As is easily seen, φ is a homomorphism of Z onto Z_n with kernel I_n. So by Theorem 4.3.3, $Z_n \simeq Z/I_n$. (This should come as no surprise, for that is how we originally introduced Z_n.)

2. Let F be a field; what can the ideals of F be? Suppose that $I \neq (0)$ is an ideal of F; let $a \neq 0 \in I$. Then, since I is an ideal of F, $1 = a^{-1}a \in I$; but now, since $1 \in I$, $r1 = r \in I$ for every $r \in F$. In short, $I = F$. So F has only the trivial ideals (0) and F itself.

3. Let R be the ring of all rational numbers having odd denominators in their reduced form. Let I be those elements of R which in reduced form have an *even* numerator; it is easy to see that I is an ideal of R. Define $\varphi: R \to Z_2$, the integers mod 2, by $\varphi(a/b) = 0$ if a is even (a, b have no common factor) and $\varphi(a/b) = 1$ if a is odd. We leave it to the reader to verify that φ is a homomorphism of R onto $\mathbb{Z}_2$ with kernel I. Thus $\mathbb{Z}_2 \simeq R/I$. Give the explicit isomorphism of R/I onto $\mathbb{Z}_2$.

4. Let R be the ring of all rational numbers whose denominators (when in reduced form) are not divisible by p, p a fixed prime. Let I be those elements in R whose numerator is divisible by p; I is an ideal of R and $R/I \simeq \mathbb{Z}_p$, the integers mod p. (Prove!)

5. Let R be the ring of all real-valued continuous functions on the closed unit interval where $(f + g)(x) = f(x) + g(x)$ and $(fg)(x) = f(x)g(x)$ for $f, g \in R$, $x \in [0, 1]$. Let $I = \{ f \in R | f(\tfrac{1}{2}) = 0 \}$. We claim that I is an ideal of R. Clearly, it is an additive subgroup. Furthermore if $f \in I$ and $g \in R$, then $f(\tfrac{1}{2}) = 0$, so $(fg)(\tfrac{1}{2}) = f(\tfrac{1}{2})g(\tfrac{1}{2}) = 0g(\tfrac{1}{2}) = 0$. Thus $fg \in I$; since I is commutative, gf is also in I. So I is an ideal of R.

What is R/I? Given any $f \in R$, then

$$f(x) = \left(f(x) - f(\tfrac{1}{2}) \right) + f(\tfrac{1}{2}) = g(x) + f(\tfrac{1}{2}),$$

where $g(x) = f(x) - f(\tfrac{1}{2})$. Because $g(\tfrac{1}{2}) = f(\tfrac{1}{2}) - f(\tfrac{1}{2}) = 0$, g is in I. So $g + I = I$. Thus $f + I = (f(\tfrac{1}{2}) + g) + I = f(\tfrac{1}{2}) + I$. Because $f(\tfrac{1}{2})$ is just a real number, R/I consists of the cosets $\alpha + I$ for α real. We claim that every real α comes up. For if $f(\tfrac{1}{2}) = \beta \neq 0$, then

$$\alpha\beta^{-1}f + I = (\alpha\beta^{-1} + I)(f + I) = (\alpha\beta^{-1} + I)(f(\tfrac{1}{2}) + I)$$

$$= (\alpha\beta^{-1} + I)(\beta + I) = \alpha\beta^{-1}\beta + I = \alpha + I.$$

So R/I consists of all $\alpha + I$, α real. Thus R/I can be shown to be isomorphic to the real field.

We now use Theorem 4.3.3 to show that $R/I \simeq$ real field via Theorem 4.3.3. Let $\varphi : R \to \mathbb{R}$ be defined by $\varphi(f) = f(\tfrac{1}{2})$. Then φ (as above) is onto and $\operatorname{Ker}\varphi = \{f \in R | f(\tfrac{1}{2}) = 0\}$; in other words, $\operatorname{Ker}\varphi = I$. So $R/I \simeq$ image of $\varphi = \mathbb{R}$.

6. Let R be the integral quaternions, in other words, $R = \{\alpha_0 + \alpha_1 i + \alpha_2 j + \alpha_3 k | \alpha_0, \alpha_1, \alpha_2, \alpha_3 \in \mathbb{Z}\}$ and let

$$I_p = \{\alpha_0 + \alpha_1 i + \alpha_2 j + \alpha_3 k \in R | \ p \, | \alpha_i \text{ for } i = 0, 1, 2, 3, \ p \text{ a fixed prime}\}.$$

The reader should verify that I_p is an ideal of R and that $R/I_p \simeq H(\mathbb{Z}_p)$ (see Problem 38 of Section 1 and the paragraph before it).

7. Let $R = \left\{ \begin{pmatrix} a & b \\ 0 & a \end{pmatrix} \middle| a, b \in \mathbb{R} \right\}$; R is a subring of the 2×2 matrices over the reals. Let $I = \left\{ \begin{pmatrix} 0 & b \\ 0 & 0 \end{pmatrix} \middle| b \in \mathbb{R} \right\}$; I is easily seen to be an additive subgroup of R. Is it an ideal of R? Consider

$$\begin{pmatrix} x & y \\ 0 & x \end{pmatrix}\begin{pmatrix} 0 & b \\ 0 & 0 \end{pmatrix} = \begin{pmatrix} 0 & xb \\ 0 & 0 \end{pmatrix};$$

so it is in I. Similarly,

$$\begin{pmatrix} 0 & b \\ 0 & 0 \end{pmatrix}\begin{pmatrix} x & y \\ 0 & x \end{pmatrix} = \begin{pmatrix} 0 & bx \\ 0 & 0 \end{pmatrix},$$

so it, too, is in I. So I is an ideal of R. What is R/I? We approach it from two points of view.

Given

$$\begin{pmatrix} a & b \\ 0 & a \end{pmatrix} \in R, \quad \text{then} \quad \begin{pmatrix} a & b \\ 0 & a \end{pmatrix} = \begin{pmatrix} a & 0 \\ 0 & a \end{pmatrix} + \begin{pmatrix} 0 & b \\ 0 & 0 \end{pmatrix}$$

so

$$\begin{pmatrix} a & b \\ 0 & a \end{pmatrix} + I = \left(\begin{pmatrix} a & 0 \\ 0 & a \end{pmatrix} + \begin{pmatrix} 0 & b \\ 0 & 0 \end{pmatrix} \right) + I = \begin{pmatrix} a & 0 \\ 0 & a \end{pmatrix} + I,$$

since $\begin{pmatrix} 0 & b \\ 0 & 0 \end{pmatrix}$ is in I. Thus all the cosets of I in R look like $\begin{pmatrix} a & 0 \\ 0 & a \end{pmatrix} + I$. If we map this onto a, that is, $\psi\left(\begin{pmatrix} a & 0 \\ 0 & a \end{pmatrix} + I \right) = a$, we can check that ψ is an isomorphism onto the real field. So $R/I \simeq \mathbb{R}$.

We see that $R/I \simeq \mathbb{R}$ another way. Define p : $R \to \mathbb{R}$ by $\varphi \begin{pmatrix} a & b \\ 0 & a \end{pmatrix}$ $= a$. We claim that φ is a homomorphism. For, given $\begin{pmatrix} a & b \\ 0 & a \end{pmatrix}$, $\begin{pmatrix} c & d \\ 0 & c \end{pmatrix}$, then

$$\varphi \begin{pmatrix} a & b \\ 0 & a \end{pmatrix} = a, \quad \varphi \begin{pmatrix} c & d \\ 0 & c \end{pmatrix} = c,$$

$$\begin{pmatrix} a & b \\ 0 & a \end{pmatrix} + \begin{pmatrix} c & d \\ 0 & c \end{pmatrix} = \begin{pmatrix} a + c & b + d \\ 0 & a + c \end{pmatrix}.$$

and

$$\begin{pmatrix} a & b \\ 0 & a \end{pmatrix}\begin{pmatrix} c & d \\ 0 & c \end{pmatrix} = \begin{pmatrix} ac & ad + bc \\ 0 & ac \end{pmatrix};$$

hence

$$\varphi\left(\begin{pmatrix} a & b \\ 0 & a \end{pmatrix} + \begin{pmatrix} c & d \\ 0 & c \end{pmatrix} \right) = \varphi \begin{pmatrix} a + c & b + d \\ 0 & a + c \end{pmatrix}$$

$$= a + c = \varphi \begin{pmatrix} a & b \\ 0 & a \end{pmatrix} + \varphi \begin{pmatrix} c & d \\ 0 & c \end{pmatrix}$$

and

$$\varphi\left(\begin{pmatrix} a & b \\ 0 & a \end{pmatrix}\begin{pmatrix} c & d \\ 0 & c \end{pmatrix} \right) = \varphi \begin{pmatrix} ac & ad + bc \\ 0 & ac \end{pmatrix}$$

$$= ac = \varphi \begin{pmatrix} a & b \\ 0 & a \end{pmatrix}\varphi \begin{pmatrix} c & d \\ 0 & c \end{pmatrix}.$$

So φ is indeed a homomorphism of R onto $\mathbb{R}$. What is Ker φ?

If $\begin{pmatrix} a & b \\ 0 & a \end{pmatrix} \in$ Ker φ, then $\varphi\begin{pmatrix} a & b \\ 0 & a \end{pmatrix} = a$, on the one hand. by the definition of φ, and $\varphi\begin{pmatrix} a & b \\ 0 & a \end{pmatrix} = 0$, since $\begin{pmatrix} a & b \\ 0 & a \end{pmatrix} \in$ Ker φ. Thus $a = 0$. From this we see that $I =$ Ker φ. So $R/I \simeq$ image of $\varphi = \mathbb{R}$ by Theorem 4.3.3.

8. Let $R = \left\{ \begin{pmatrix} a & b \\ -b & a \end{pmatrix} \middle| a, b \in \mathbb{R} \right\}$ and let $\mathbb{C}$ be the field of complex numbers. Define $\psi : R \to \mathbb{C}$ by $\psi\begin{pmatrix} a & b \\ -b & a \end{pmatrix} = a + bi$. We leave it to the reader to verify that ψ is an isomorphism of R onto $\mathbb{C}$. So R is isomorphic to the field of complex numbers.

9. Let R be any commutative ring with 1. If $a \in R$. let $(a) = \{ xa | x \in R \}$. We claim that (a) is an ideal of R. To see this. suppose that $u, v \in (a)$; thus $u = xa$, $v = ya$ for some $x, y \in R$. whence

$$u \pm v = xa \pm ya = (x \pm y)a \in (a).$$

Also, if $u \in (a)$ and $r \in R$, then $u = xa$, hence $ru = r(xa) = (rx)a$. so is in (a). Thus (a) is an ideal of R.

Note that if R is not commutative, then (a) need not be an ideal; but it is certainly a left ideal of R.

Problems

Easier Problems

1. If R is a commutative ring and $a \in R$. let $L(a) = \{ x \in R | xa = 0 \}$. Prove that $L(a)$ is an ideal of R.

2. If R is a commutative ring with 1 and R has no ideals other than (0) and itself, prove that R is a field. (**Hint:** Look at Example 9.)

*3. If $\varphi : R \to R'$ is a homomorphism of R *onto* R' and R has a unit element, 1, show that $\varphi(1)$ is the unit element of R'.

4. If I, J are ideals of R, define $I + J$ by $I + J = \{ i + j | i \in I. j \in J \}$. Prove that $I + J$ is an ideal of R.

5. If I is an ideal of R and A is a subring of R. show that $I \cap A$ is an ideal of A.

6. If I, J are ideals of R. show that $I \cap J$ is an ideal of R.

7. Give a complete proof of Theorem 4.3.2.

8. Give a complete proof of Theorem 4.3.4.

9. Let $\varphi: R \rightarrow R'$ be a homomorphism of R onto R' with kernel K. If A' is a subring of R', let $A = \{ a \in R | \varphi(a) \in A' \}$. Show that:
   (a) A is a subring of R, $A \supset K$.
   (b) $A/K \simeq A'$.
   (c) If A' is a left ideal of R', then A is a left ideal of R.

10. Prove Theorem 4.3.5.

11. In Example 3, give the explicit isomorphism of R/I onto $\mathbb{Z}_2$.

12. In Example 4, show that $R/I \simeq \mathbb{Z}_p$.

13. In Example 6, show that $R/I_p \simeq H(\mathbb{Z}_p)$.

14. In Example 8, verify that the mapping ψ given is an isomorphism of R onto $\mathbb{C}$.

15. If I, J are ideals of R, let IJ be the set of all *sums of elements of the form ij*, where $i \in I, j \in J$. Prove that IJ is an ideal of R.

16. Show that the ring of 2×2 matrices over the reals has nontrivial left ideals (and also nontrivial right ideals).

17. If A is a subring of R and I is an ideal of R. let

$$A + I = \{ a + i | a \in A, i \in I \}.$$

Prove that:
   (a) $A + I$ is a subring of R. $A + I \supset I$.
   (b) $(A + I)/I \simeq A/(A \cap I)$.

   If R, S are rings, define the *direct sum* of R and S, $R \oplus S$, by

$$R \oplus S = \{ (r, s) | r \in R, s \in S \}$$

where $(r, s) = (r_1, s_1)$ if and only if $r = r_1, s = s_1$, and where

$$(r, s) + (t, u) = (r + t, s + u), \qquad (r, s)(t, u) = (rs, tu).$$

18. Show that $R \oplus S$ is a ring and that the subrings $\{(r.0)|r \in R\}$ and $\{(0, s)|s \in S\}$ are ideals of $R \oplus S$ isomorphic to R and S, respectively.

19. If $R = \left\{ \begin{pmatrix} a & b \\ 0 & c \end{pmatrix} \middle| a, b, c \text{ real} \right\}$ and $I = \left\{ \begin{pmatrix} 0 & b \\ 0 & 0 \end{pmatrix} \middle| b \text{ real} \right\}$, show that:
   (a) R is a ring.
   (b) I is an ideal of R.
   (c) $R/I \simeq F \oplus F$, where F is the field of real numbers.

20. If I, J are ideals of R, let $R_1 = R/I$ and $R_2 = R/J$. Show that $\varphi: R \to R_1 \oplus R_2$ defined by $\varphi(r) = (r + I, r + J)$ is a homomorphism of R into $R_1 \oplus R_2$ such that $\operatorname{Ker} \varphi = I \cap J$.

21. Let $\mathbb{Z}_{15}$ be the ring of integers mod 15. Show that $\mathbb{Z}_{15} \simeq \mathbb{Z}_3 \oplus \mathbb{Z}_5$.

Middle-Level Problems

22. Let $\mathbb{Z}$ be the ring of integers and m, n two relatively prime integers, I_m the multiples of m in $\mathbb{Z}$, and I_n the multiples of n in $\mathbb{Z}$.
   (a) What is $I_m \cap I_n$?
   (b) Use the result of Problem 20 to show that there is an isomorphism of $\mathbb{Z}/I_{mn}$ into $\mathbb{Z}/I_m \oplus \mathbb{Z}/I_n$.
   (c) By counting elements on both sides, show that
$$\mathbb{Z}/I_{mn} \simeq \mathbb{Z}_m \oplus \mathbb{Z}_n.$$

23. If m, n are relatively prime, prove that $\mathbb{Z}_{mn} \simeq \mathbb{Z}_m \oplus \mathbb{Z}_n$. (**Hint:** Problem 22.)

*24. Use the result of Problem 22 or Problem 23 to prove the *Chinese Remainder Theorem*, which asserts that if m and n are relatively prime integers and a, b any integers, we can find an integer x such that $x \equiv a \bmod m$ and $x \equiv b \bmod n$ simultaneously.

25. Let R be the ring of 2×2 matrices over the real numbers; suppose that I is an ideal of R. Show that $I = (0)$ or $I = R$. (Contrast this with the result of Problem 16.)

Harder Problems

26. Let R be a ring with 1 and let S be the ring of 2×2 matrices over R. If I is an ideal of S show that there is an ideal J of R such that I consists of all the 2×2 matrices over J.

27. If $p_1, p_2, \ldots, p_n$ are distinct primes, show that there are exactly 2^n solutions of $x^2 \equiv x \bmod (p_1 \cdots p_n)$, where $0 \le x < p_1 \cdots p_n$.

28. Suppose that R is a ring whose only left ideals are (0) and R itself. Prove that either R is a division ring or R has p elements. p a prime, and $ab = 0$ for every $a, b \in R$.

29. Let R be a ring with 1. An element $a \in R$ is said to have a *left inverse* if $ba = 1$ for some $b \in R$. Show that if the left inverse b of a is unique, then $ab = 1$ (so b is also a right inverse of a).

4.   MAXIMAL IDEALS

This will be a section with one major theorem. The importance of this result will only become fully transparent when we discuss fields in Chapter 5. However, it is a result that stands on its own two feet. It isn't difficult to prove, but in mathematics the correlation between difficult and important isn't always that high. There are many difficult results that are of very little interest and of even less importance, and some easy results that are crucial. Of course, there are some results—many, many —which are of incredible difficulty and importance.

> **Lemma 4.4.1.**   Let R be a commutative ring with unit whose only ideals are (0) and itself. Then R is a field.

Proof.   Let $a \ne 0$ be in R. Then $(a) = \{xa \mid x \in R\}$ is an ideal of R. as we verified in Example 9 in the preceding section. Since $a = 1a \in (a)$, $(a) \ne 0$. Thus, by our hypothesis on R, $(a) = R$. But then, by the definition of (a), every element $i \in R$ is a multiple xa of a for some $x \in R$. In particular, because $1 \in R$, $1 = ba$ for some $b \in R$. This shows that a has the inverse b in R. So R is a field. $\square$

In Theorem 4.3.4—the Second Homomorphism Theorem—we saw that if $\varphi: R \to R'$ is a homomorphism of R onto R' with kernel K, then there is a 1-1 correspondence between ideals of R' and ideals of R that contain K. Suppose that there are no ideals other than K itself and R which contain K. What does this imply about R'? Since (0) in R' corresponds to K in R, and R corresponds to R' in this correspondence given by the Second Homomorphism Theorem, we must conclude that in this case R' has no ideals other than (0) and itself. So if R' is commutative and has a unit element, then, by Lemma 4.4.1, R' must be a field.

This prompts the following definition.

Definition. A proper ideal M of R is a *maximal ideal* of R if the only ideals of R that contain M are M itself and R.

The discussion preceding this definition has already almost proved for us

Theorem 4.4.2. Let R be a commutative ring with 1, and let M be a maximal ideal of R. Then R/M is a field.

Proof. There is a homomorphism of R onto $R' = R/M$, and since $1 \in R$ we have that R' has $1 + M$ as its unit element. (See Problem 3, Section 3). Because M is a maximal ideal of R, we saw in the discussion above that R' has no nontrivial ideals. Thus, by Lemma 4.4.1, $R' = R/M$ is a field.

This theorem will be our entry into the discussion of fields, for it will enable us to construct particularly desirable fields whenever we shall need them.

Theorem 4.4.2 has a converse. This is

Theorem 4.4.3. If R is a commutative ring with 1 and M an ideal of R such that R/M is a field, then M is a maximal ideal of R.

Proof. We saw in Example 2 of Section 3 that the only ideals in a field F are (0) and F itself. Since R/M is a field, it has only (0) and itself as ideals. But then, by the correspondence given us by Theorem 4.3.4, there can be no ideal of R properly between M and R. Thus M is a maximal ideal of R. $\square$

We give a few examples of maximal ideal in commutative rings.

EXAMPLES

1. Let $\mathbb{Z}$ be the integers and M an ideal of $\mathbb{Z}$. As an ideal of $\mathbb{Z}$ we certainly have that M is an additive subgroup of $\mathbb{Z}$, so must consist of all multiples of some fixed integer n. Thus since $R/M \simeq \mathbb{Z}_n$ and since $\mathbb{Z}_n$ is a field if and only if n is a prime, we see that M is a maximal ideal of $\mathbb{Z}$ if and only if M consists of all the multiples of some prime p. Thus the set of maximal ideals in $\mathbb{Z}$ corresponds to the set of prime numbers.

2. Let $\mathbb{Z}$ be the integers, and let $R = \{a + bi | a, b \in \mathbb{Z}\}$, a subring of $\mathbb{C}$ ($i^2 = -1$). In R let M be the set of all $a + bi$ in R, where $3|a$ and $3|b$. We leave it to the reader to verify that M is an ideal of R.

We claim that M is a maximal ideal of R. For suppose that $N \supset M$ and $N \neq M$ is an ideal of R. So there is an element $r + si \in N$, where 3 *divides neither r nor s*. Therefore, $3 \nmid (r^2 + s^2)$. (Prove!) But $t = r^2 + s^2 = (r + si)(r - si)$, so is in N, since $r + si \in N$ and N is an ideal of R. So N has an integer $t = r^2 + s^2$ *not* divisible by 3. Thus $ut + 3v = 1$ for some integers u, v; but $t \in N$, hence $ut \in N$ and $3 \in M \subset N$, so $3v \in N$. Therefore, $1 = ut + 3v \in N$. Therefore, $(a + bi)1 \in N$, since N is an ideal of R, for all $a + bi \in R$. This tells us that $N = R$. So the only ideal of R above M is R itself. Consequently, M is a maximal ideal of R.

By Theorem 4.4.2 we know that R/M is a field. It can be shown (see Problem 2) that R/M is a field having nine elements.

3. Let R be as in Example 2 and let $I = \{a + bi | 5|a$ and $5|b\}$. We assert that I is *not* a maximal ideal of R.

In R we can factor $5 = (2 + i)(2 - i)$. Let $M = \{x(2 + i) | x \in R\}$. M is an ideal of R, and since $5 = (2 + i)(2 - i)$ is in M, we see that $I \subset M$. Clearly, $I \neq M$ for $2 + i \in M$ and is not in I because $5 \nmid 2$. So $I \neq M$. Can $M = R$? If so, then $(2 + i)(a + bi) = 1$ for some a, b. This gives $2a - b = 1$ and $2b + a = 0$; these two equations imply that $5a = 2$, so $a = \frac{2}{5}$, $b = -\frac{1}{5}$. But $\frac{2}{5} \notin \mathbb{Z}$, $-\frac{1}{5} \notin \mathbb{Z}$; the element $a + bi = \frac{2}{5} - \frac{1}{5}i$ is *not* in R. So $M \neq R$.

One can show, however, that M is a maximal ideal of R. (See Problem 3.)

4. Let $R = \{a + b\sqrt{2} | a, b$ integers$\}$, which is a subring of the real field under the sum and product of real numbers. That R is a ring follows from

$$(a + b\sqrt{2}) + (c + d\sqrt{2}) = (a + c) + (b + d)\sqrt{2}$$

and

$$(a + b\sqrt{2})(c + d\sqrt{2}) = (ac + 2bd) + (ad + bc)\sqrt{2}.$$

Let $M = \{a + b\sqrt{2} \in R \mid 5|a \text{ and } 5|b\}$. M is easily seen to be an ideal of R. We leave it to the reader to show that M is a maximal ideal of R and that R/M is a field having 25 elements.

5. Let R be the ring of all real-valued continuous functions on the closed unit interval $[0, 1]$. We showed in Example 5 of Section 3 that if $M = \{f \in R \mid f(\frac{1}{2}) = 0\}$, then M is an ideal of R and R/M is isomorphic to the real field. Thus, by Theorem 4.4.3, M is a maximal ideal of R.

Of course, if we let $M_\gamma = \{f \in R \mid f(\gamma) = 0\}$, where $\gamma \in [0, 1]$, then M_γ is also a maximal ideal. It can be shown that every maximal ideal in R is of the form M_γ for some $\gamma \in [0, 1]$, but to prove it we would require some results from real variable theory.

What this example shows is that the maximal ideals in R correspond to the points of $[0, 1]$.

Problems

1. If a, b are integers and $3 \nmid a$, $3 \nmid b$, show that $3 \nmid (a^2 + b^2)$.

2. Show that in Example 2, R/M is a field having nine elements.

3. In Example 3, show that $M = \{x(2 + i) \mid x \in R\}$ is a maximal ideal of R.

4. In Example 3, show that $R/M \simeq \mathbb{Z}_5$.

5. In Example 3, shows that $R/I \simeq \mathbb{Z}_5 \oplus \mathbb{Z}_5$.

6. In Example 4, show that M is a maximal ideal of R.

7. In Example 4, show that R/M is a field having 25 elements.

8. Using Example 2 as a model, construct a field having 49 elements.

We make a short excursion back to congruences mod p, where p is an *odd* prime. If a is an integer such that $p \nmid a$ and $x^2 \equiv a \bmod p$ has a solution for x in Z, we say that a is a *quadratic residue* mod p. Otherwise, a is said to be a *quadratic nonresidue* mod p.

9. Show that $(p - 1)/2$ of the numbers $1, 2, \ldots, p - 1$ are quadratic residues and $(p - 1)/2$ are quadratic nonresidues mod p. [**Hint:** Show that $\{x^2 | x \neq 0 \in Z_p\}$ forms a group of order $(p - 1)/2$.]

10. Let $m > 0$ be in Z, and suppose that m is *not* a square in Z. Let $R = \{a + \sqrt{m}\, b | a, b \in Z\}$ Prove that under the operations of sum and product of real numbers R is a ring.

11. If p is an odd prime, let the set $I_p = \{a + \sqrt{m}\, b : p|a \text{ and } p|b\}$, where $a + \sqrt{m}\, b \in R$, the ring in Problem 10. Show that I_p is an ideal of R.

12. If m is a *quadratic nonresidue* mod p, show that the ideal I_p in Problem 11 is a maximal ideal of R.

13. In Problem 12 show that R/I_p is a field having p^2 elements.

5. POLYNOMIAL RINGS

The material that we consider in this section involves the notion of polynomial and the set of all polynomials over a given field. We hope that most readers will have some familiarity with the notion of polynomial from their high school days and will have seen some of the things one does with polynomials: factoring them, looking for their roots, dividing one by another to get a remainder, and so on. The emphasis we shall give to the concept and algebraic object known as a polynomial ring will be in a quite different direction from that given in high school.

Be that as it may, what we shall strive to do here is to introduce the ring of polynomials over a field and show that this ring is amenable to a careful dissection that reveals its innermost structure. As we shall see, this ring is very well-behaved. The development should remind us of what was done for the ring of integers in Section 5 of Chapter 1. Thus we shall run into the analog of Euclid's algorithm, greatest common divisor, divisibility, and possibly most important, the appropriate analog of prime number. This will lead to unique factorization of a general polynomial into these "prime polynomials," and to the nature of the ideals and the maximal ideals in this new setting.

But the polynomial ring enjoys one feature that the ring of integers did not: the notion of a root of a polynomial. The study of the nature of such roots—which will be done, for the most part, in the next chapter—constitutes a large and important part of the algebraic history of the past. It goes under the title *Theory of Equations*, and in its honorable past, a large variety of magnificent results have been obtained in this area. Hopefully, we shall see some of these as our development progresses.

With this sketchy outline of what we intend to do out of the way, we now get down to the nitty-gritty of doing it.

Let F be a field. *By the ring of polynomials in x over F*, which we shall always write as $F[x]$, we mean the set of all formal expressions $p(x) = a_0 + a_1x + \cdots + a_{n-1}x^{n-1} + a_nx^n, n \geq 0$, where the a_i, the *coefficients* of the *polynomial* $p(x)$, are in F. In $F[x]$ we define equality, sum, and product of two polynomials so as to make of $F[x]$ a commutative ring as follows:

1. Equality. We declare $p(x) = a_0 + a_1x + \cdots + a_nx^n$ and $q(x) = b_0 + b_1x + \cdots + b_nx^n$ to be *equal* if and only if their corresponding coefficients are equal, that is, if and only if $a_i = b_i$ for all $i \geq 0$.

We combine this definition of equality of polynomials $p(x)$ and $q(x)$ with the convention that if

$$q(x) = b_0 + b_1x + \ldots + b_nx^n$$

and if $b_{m+1} = \ldots = b_n = 0$, then we can drop the last $n - m$ terms and write $q(x)$ as

$$q(x) = b_0 + b_1x + \ldots b_mx^m.$$

This convention is followed in the definition of addition that follows, where s is the larger of m and n and we add coefficients $a_{n+1} = \ldots = a_s = 0$ if $n < s$ or $b_{m+1} = \ldots = b_s = 0$ if $m < s$.

2. Addition. If $p(x) = a_0 + a_1x + \cdots + a_nx^n$ and $q(x) = b_0 + b_1x + \cdots b_mx^m$, we declare $p(x) + q(x) = c_0 + c_1x + \cdots c_sx^s$, where for each i, $c_i = a_i + b_i$.

So we add polynomials by adding their corresponding coefficients.

The definition of multiplication is a little more complicated. We define it loosely at first and then more precisely.

3.   Multiplication.   If $p(x) = a_0 + a_1x + \cdots + a_nx^n$ and $q(x) = b_0 + b_1x + \cdots + b_mx^m$, we declare $p(x)q(x) = c_0 + c_1x + \cdots + c_tx^t$, where the c_i are determined by multiplying the expression out formally, using the distributive laws and the rules of exponents $x^ux^v = x^{u+v}$, and collecting terms. More formally,

$$c_i = a_ib_0 + a_{i-1}b_1 + \cdots + a_1b_{i-1} + a_0b_i, \qquad \text{for every } i.$$

We illustrate these operations with a simple example, but first a notational device: If some coefficient is 0, we just omit that term. Thus we write $9 + 0x + 7x^2 + 0x^3 - 14x^4$ as $9 + 7x^2 - 14x^4$.

Let $p(x) = 1 + 3x^2$, $q(x) = 4 - 5x + 7x^2 - x^3$. Then $p(x) + q(x) = 5 - 5x + 10x^2 - x^3$ while

$$
\begin{aligned}
p(x)q(x) &= (1 + 3x^2)(4 - 5x + 7x^2 - x^3) \\
&= 4 - 5x + 7x^2 - x^3 + 3x^2(4 - 5x + 7x^2 - x^3) \\
&= 4 - 5x + 7x^2 - x^3 + 12x^2 - 15x^3 + 21x^4 - 3x^5 \\
&= 4 - 5x + 19x^2 - 16x^3 + 21x^4 - 3x^5.
\end{aligned}
$$

Try this product using the c_i as given above.

In some sense this definition of $F[x]$ is not a definition at all. We have indulged in some hand waving in it. But it will do. We could employ sequences to formally define $F[x]$ more precisely, but it would merely cloud what to most readers is well known.

The first remark that we make—and do not verify—is that $F[x]$ is a commutative ring. To go through the details of checking the axioms for a commutative ring is a straightforward but laborious task. However, it is important to note

Lemma 4.5.1. $F[x]$ is a commutative ring with unit.

Definition.   If $p(x) = a_0 + a_1x + \cdots a_nx^n$ and $a_n \neq 0$, then the *degree* of $p(x)$, denoted by $\deg p(x)$, is n.

So the degree of a polynomial $p(x)$ is the highest power of x that occurs in the expression for $p(x)$ with a nonzero coefficient. Thus

$\deg(x - x^2 + x^4) = 4$, $\deg(7x) = 1$, $\deg 7 = 0$. (Note that this definition does not assign a degree to 0. It is, however, sometimes convenient to adopt the convention that the degree of 0 be $-\infty$, in which case many degree related results will hold in this extended context.) The polynomials of degree 0 are called the *constants;* thus the set of constants can be identified with F.

The degree function on $F[x]$ will play a similar role to that played by the size of the integers in Z, in that it will provide us with a Euclid algorithm for $F[x]$.

One immediate and important property of the degree function is that it behaves well for products.

Lemma 4.5.2. If $p(x)$, $q(x)$ are nonzero elements of $F[x]$, then $\deg(p(x)q(x)) = \deg p(x) + \deg q(x)$.

Proof. Let $m = \deg p(x)$ and $n = \deg q(x)$; thus the polynomial $p(x) = a_0 + a_1 x + \cdots + a_m x^m$, where $a_m \neq 0$, and the polynomial $q(x) = b_0 + b_1 x + \cdots + b_n x^n$, where $b_n \neq 0$. The highest power of x that can occur in $p(x)q(x)$ is x^{m+n}, from our definition of the product. What is the coefficient of x^{m+n}? The only way that x^{m+n} can occur is from $(a_m x^m)(b_n x^n) = a_m b_n x^{m+n}$. So the coefficient of x^{m+n} in $p(x)q(x)$ is $a_m b_n$, which is not 0, since $a_m \neq 0$, $b_n \neq 0$. Thus $\deg(p(x)q(x)) = m + n = \deg p(x) + \deg q(x)$, as claimed in the lemma. $\square$

One also has some information about $\deg(p(x) + q(x))$. This is

Lemma 4.5.3. If $p(x), q(x) \in F[x]$ and $p(x) + q(x) \neq 0$, then $\deg(p(x) + q(x)) \leq \max(\deg p(x), \deg q(x))$.

We leave the proof of Lemma 4.5.3 to the reader. It will play no role in what is to come, whereas Lemma 4.5.2 will be important. We put it in so that the "+" should not feel slighted vis-à-vis the product.

An immediate consequence of Lemma 4.5.2 is

Lemma 4.5.4. $F[x]$ is an integral domain.

Proof.   If $p(x) \neq 0$ and $q(x) \neq 0$, then $\deg p(x) \geq 0$, $\deg q(x) \geq 0$, so $\deg(p(x)q(x)) = \deg p(x) + \deg q(x) \geq 0$. Therefore, $p(x)q(x)$ has a degree, so cannot be 0 (which has no degree assigned to it). Thus $F[x]$ is an integral domain. $\square$

One of the things that we were once forced to learn was to divide one polynomial by another. How did we do this? The process was called *long division*. We illustrate with an example how this was done, for what we do in the example is the model of what we shall do in the general case.

We want to divide $2x^2 + 1$ $x^4 - 7x + 1$. We do it schematically as follows:

$$
\begin{array}{r}
\frac{1}{2}x^2 - \frac{1}{4}x \\
(2x^2 + 1)\overline{)\,x^4 \qquad\qquad - 7x + 1\,} \\
\underline{x^4 + \frac{1}{2}x^2} \\
-\frac{1}{2}x^2 \;-\; 7x + 1 \\
\underline{-\frac{1}{2}x^2 \qquad\quad -\frac{1}{4}} \\
- 7x + 1\tfrac{1}{4}
\end{array}
$$

and we interpret this as:

$$x^4 - 7x + 1 = (2x^2 + 1)\left(\tfrac{1}{2}x^2 - \tfrac{1}{4}\right) + \left(-7x + \tfrac{5}{4}\right)$$

and $-\lfloor 7x \rfloor + \tfrac{5}{4}$ is called the *remainder* in this division.

What exactly did we do? First, where did the $\frac{1}{2}x^2$ come from? It came from the fact that when we multiply $2x^2 + 1$ by $\frac{1}{2}x^2$ we get x^4, the highest power occurring in $x^4 - 7x + 1$. So subtracting $\frac{1}{2}x^2(2x^2 + 1)$ from $x^4 - 7x + 1$ gets rid of the x^4 term and we go on to what is left and repeat the procedure.

This "repeat the procedure" suggests induction, and that is how we shall carry out the proof. But keep in mind that all we shall be doing is what we did in the example above.

What this gives us is something like Euclid's Algorithm, in the integers. However, here we call it the *Division Algorithm*.

Theorem 4.5.5 (Division Algorithm).   Given the polynomials $f(x)$, $g(x) \in F[x]$, where $g(x) \neq 0$, then

$$f(x) = q(x)g(x) + r(x),$$

where $q(x), r(x) \in F[x]$ and $r(x) = 0$ or $\deg r(x) < \deg g(x)$.

Proof. We go by induction on $\deg f(x)$. If either $f(x) = 0$ or $\deg f(x) < \deg g(x)$, then $f(x) = 0g(x) + f(x)$, which satisfies the conclusion of the theorem.

So suppose that $\deg f(x) \geq \deg g(x)$; thus the polynomial $f(x) = a_0 + a_1 x + \cdots + a_m x^m$, where $a_m \neq 0$ and the polynomial $g(x) = b_0 + b_1 x + \cdots + b_n x^n$, where $b_n \neq 0$ and where $m \geq n$.

Consider

$$\frac{a_m}{b_n} x^{m-n} g(x) = \frac{a_m}{b_n} x^{m-n} (b_0 + b_1 x + \cdots + b_n x^n)$$

$$= \frac{a_m b_0}{b_n} x^{m-n} + \cdots + a_m x^m;$$

thus $(a_m/b_n) x^{m-n} g(x)$ *has the same degree and same highest coefficient as does* $f(x)$, *so* $f(x) - (a_m/b_n) x^{m-n} g(x) = h(x)$ is such that the relation $\deg h(x) < \deg f(x)$ holds. Thus, by induction,

$$h(x) = q_1(x) g(x) + r(x), \qquad \text{where} \qquad q_1(x), r(x) \in F[x]$$

and $r(x) = 0$ or $\deg r(x) < \deg g(x)$. Remembering what $h(x)$ is, we get

$$h(x) = f(x) - \frac{a_m}{b_n} x^{m-n} g(x) = q_1(x) g(x) + r(x)$$

so

$$f(x) = \left(\frac{a_m}{b_n} x^{m-n} + q_1(x) \right) g(x) + r(x).$$

If $q(x) = (a_m/b_n) x^{m-1} + q_1(x)$, we have achieved the form claimed in the theorem. $\square$

The Division Algorithm has one immediate application: It allows us to determine the nature of all the ideals of $F[x]$. As we see in the next theorem, an ideal of $F[x]$ must merely consist of all multiples, by elements of $F[x]$, of some *fixed* polynomial.

Theorem 4.5.6. If $I \neq (0)$ is an ideal of $F[x]$, then $I = \{f(x)g(x)|f(x) \in F[x]\}$; that is, I consists of all multiples of the fixed polynomial $g(x)$ by the elements of $F[x]$.

Proof. To prove the theorem, we need to produce that fixed polynomial $g(x)$. Where are we going to dig it up? The one control we have numerically on a given polynomial is its degree. So why not use the degree function as the mechanism for finding $g(x)$.

Since $I \neq (0)$ there are elements in I having nonnegative degree. So there is a polynomial $g(x) \neq 0$ in I of minimal degree; that is, $g(x) \neq 0$ is in I and if $0 \neq t(x) \in I$, then $\deg t(x) \geq \deg g(x)$. Thus, by the division algorithm, $t(x) = q(x)g(x) + r(x)$, where $r(x) = 0$ or $\deg r(x) < \deg g(x)$. But since $g(x) \in I$ and I is an ideal of $F[x]$, we have that $q(x)g(x) \in I$. By assumption, $t(x) \in I$, thus $t(x) - q(x)g(x)$ is in I, so $r(x) = t(x) - q(x)g(x)$ is in I. Since $g(x)$ has minimal degree for the elements of I and $r(x) \in I$, $\deg r(x)$ *cannot* be less than $\deg g(x)$. So we are left with $r(x) = 0$. But this says that $t(x) = q(x)g(x)$. So every element in I is a multiple of $g(x)$. On the other hand, since $g(x) \in I$ and I is an ideal of $F[x]$, $f(x)g(x) \in I$ for all $f(x) \in F[x]$. The net result of all this is that $I = \{f(x)g(x)|f(x) \in F[x]\}$. $\square$

Definition. An integral domain R is called a *principal ideal domain* if every ideal I in R is of the form $I = \{xa|x \in R\}$ for some $a \in I$.

Theorem 4.5.6 can be stated as: $F[x]$ *is a principal ideal domain*.

We shall write the *ideal generated by a given polynomial*, $g(x)$, namely $\{f(x)g(x)|f(x) \in F[x]\}$, as $(g(x))$.

The proof showed that if I is an ideal of R, then $I = (g(x))$, where $g(x)$ is a polynomial of lowest degree contained in I. But $g(x)$ is not unique, for if $a \neq 0 \in F$, then $ag(x)$ is in I and has the same degree as $g(x)$, so $I = (ag(x))$.

To get some sort of uniqueness in all this, we single out a class of polynomials.

Definition. $f(x) \in F[x]$ is a *monic polynomial* if the coefficient of its highest power is 1.

Thus $f(x)$ is monic means that

$$f(x) = x^n + a_{n-1}x^{n-1} + \cdots + a_1x + a_0.$$

We leave to the reader to show that if I is an ideal of $F[x]$, then there is *only one* monic polynomial of lowest degree in I. Singling this out as the generator of I does give us a "monic" uniqueness for the generation of I.

Our next step in this parallel development with what happens in the integers is to have the notion of one polynomial dividing another.

> **Definition.** If $f(x)$ and $g(x) \neq 0 \in F[x]$, then $g(x)$ divides $f(x)$, written as $g(x)|f(x)$, if $f(x) = a(x)g(x)$ for some $a(x) \in F[x]$.

Note that if $g(x)|f(x)$, then $\deg g(x) \leq \deg f(x)$ by Lemma 4.5.2. Note, too, that if $g(x)|f(x)$, then the ideals $(f(x))$ and $(g(x))$ of $F[x]$, generated by $f(x)$ and $g(x)$, respectively, satisfy the containing relation $(f(x)) \subset (g(x))$. (Prove!)

We again emphasize the parallelism between Z, the integers, and $F[x]$ by turning to the notion of *greatest common divisor*. In order to get some sort of uniqueness, we shall insist that the greatest common divisor always be a monic polynomial.

> **Definition.** For any two polynomials $f(x)$ and $g(x) \in F[x]$ (not both 0,) the polynomial $d(x) \in F[x]$ is the *greatest common divisor* of $f(x)$, $g(x)$ if $d(x)$ is a monic polynomial such that:
> (a) $d(x)|f(x)$ and $d(x)|g(x)$.
> (b) If $h(x)|f(x)$ and $h(x)|g(x)$, then $h(x)|d(x)$.

Although we defined the greatest common divisor of two polynomials, we neither know, as yet, that it exists, nor what its form may be. We could define it in another, and equivalent, way as the *monic polynomial of highest degree that divides both $f(x)$ and $g(x)$*. If we did that, its existence would be automatic, but we would not know its form.

Theorem 4.5.7. Given $f(x)$ and $g(x) \neq 0$ in $F[x]$, then their greatest common divisor $d(x) \in F[x]$ exists; moreover, $d(x) = a(x)f(x) + b(x)g(x)$ for some $a(x), b(x) \in F[x]$.

Proof. Let I be the set of all $r(x)f(x) + s(x)g(x)$ as $r(x), s(x)$ run freely over $F[x]$. We claim that I is an ideal of R. For,

$$(r_1(x)f(x) + s_1(x)g(x)) + (r_2(x)f(x) + s_2(x)g(x))$$
$$= (r_1(x) + r_2(x))f(x) + (s_1(x) + s_2(x))g(x),$$

so is again in I, and for $t(x) \in F[x]$,

$$t(x)(r(x)f(x) + s(x)g(x)) = (t(x)r(x))f(x) + (t(x)s(x))g(x),$$

so it, too, is again in I. Thus I is an ideal of $F[x]$. Since $g(x) \neq 0$, we know that $I \neq 0$, since both $f(x)$ and $g(x)$ are in I.

Since $I \neq 0$ is an ideal of $F[x]$, it is generated by a *unique monic* polynomial $d(x)$ (Theorem 4.5.6). Since $f(x), g(x)$ are in I, they must then be multiples of $d(x)$ by elements of $F[x]$. This assures us that $d(x)|f(x)$ and $d(x)|g(x)$.

Because $d(x) \in I$ and I is of the set of all $r(x)f(x) + s(x)g(x)$, we have that $d(x) = a(x)f(x) + b(x)g(x)$ for some appropriate $a(x), b(x) \in F[x]$. Thus if $h(x)|f(x)$ and $h(x)|g(x)$, then $h(x)|(a(x)f(x) + b(x)g(x) = d(x)$. So $d(x)$ is the greatest common divisor of $f(x)$ and $g(x)$.

This proves the theorem; the uniqueness of $d(x)$ is guaranteed by the demand that we have made that the greatest common divisor be monic. $\square$

Another way to see the uniqueness of $d(x)$ is from

Lemma 4.5.8. If $f(x) \neq 0$, $g(x) \neq 0$ are in $F[x]$ and $f(x)|g(x)$ and $g(x)|f(x)$, then $f(x) = ag(x)$, where $a \in F$.

Proof. By the mutual divisibility condition on $f(x)$ and $g(x)$ we have, by Lemma 4.5.2, $\deg f(x) \leq \deg g(x) \leq \deg f(x)$, so $\deg f(x) = \deg g(x)$. But $f(x) = a(x)g(x)$, so

$$\deg f(x) = \deg a(x) + \deg g(x) = \deg a(x) + \deg f(x),$$

in consequence of which $\deg a(x) = 0$, so $a(x) = a$, an element of F. $\square$

We leave the proof of the uniqueness of the greatest common divisor via Lemma 4.5.8 to the reader.

Definition. The polynomials $f(x), g(x)$ in $F[x]$ are said to be *relatively prime* if their greatest common divisor is 1.

Although it is merely a very special case of Theorem 4.5.7, to emphasize it and to have it to refer to, we state:

Theorem 4.5.9. If $f(x), g(x) \in F[x]$ are relatively prime, then $a(x)f(x) + b(x)g(x) = 1$ for some $a(x), b(x) \in F[x]$. Conversely, if $a(x)f(x) + b(x)g(x) = 1$ for some $a(x)$, $b(x) \in F[x]$, then $f(x)$ and $g(x)$ are relatively prime.

Proof. We leave this "conversely" part to the reader as exercise. $\square$

As with the integers, we have

Theorem 4.5.10. If $q(x)$ and $f(x)$ are relatively prime and if $q(x)|f(x)g(x)$, then $q(x)|g(x)$.

Proof. By Theorem 4.5.9 $a(x)f(x) + b(x)q(x) = 1$ for some $a(x)$, $b(x) \in F[x]$. Therefore,

$$(1) \qquad a(x)f(x)g(x) + b(x)q(x)g(x) = g(x).$$

Since $q(x)|b(x)g(x)q(x)$ and $q(x)|f(x)g(x)$ by hypothesis, $q(x)$ divides the left-hand side of the relation in (1). Thus $q(x)$ divides the right-hand side of (1), that is, $q(x)|g(x)$, the desired conclusion. $\square$

We are now ready to single out the important class of polynomials that will play the same role as prime objects in $F[x]$ as did the prime numbers in Z.

Definition. The polynomial $p(x) \in F[x]$ of positive degree is *irreducible* in $F[x]$ if, given any polynomial $f(x)$ in $F[x]$, then either $p(x)|f(x)$ or $p(x)$ is relatively prime to $f(x)$.

From the definition, it follows that $p(x)$ is irreducible in $F[x]$ if and only if $P(x)$ cannot be factored as a product of two polynomials of positive degree. (Prove!) That is, in other words, if $p(x) = a(x)b(x)$, where $a(x)$ and $b(x)$ are in $F[x]$, then either $a(x)$ is a constant or $b(x)$ is a constant (constant = element of F).

Note that the irreducibility of a polynomial depends on the field F. For instance, the polynomial $x^2 - 2$ is irreducible in $Q[x]$, where Q is the field of rational numbers, but $x^2 - 2$ is not irreducible in $\mathbb{R}[x]$, where $\mathbb{R}$ is the field of real numbers, for in $\mathbb{R}[x]$

$$x^2 - 2 = (x - \sqrt{2})(x + \sqrt{2}).$$

Corollary to Theorem 4.5.10. If $p(x)$ is irreducible in $F[x]$ and $p(x) | a_1(x)a_2(x) \cdots a_k(x)$, where $a_1(x), \ldots, a_k(x)$ are in $F[x]$, then $p(x) | a_i(x)$ for some i.

Proof. We leave the proof to the reader. (See Theorem 1.5.6.) $\square$

Aside from its other properties, an irreducible polynomial $p(x)$ in $F[x]$ enjoys the property that $(p(x))$, the ideal generated by $p(x)$ in $F[x]$, is a maximal ideal of $F[x]$. We prove this now.

Theorem 4.5.11. If $p(x) \in F[x]$, then the ideal $(p(x))$ generated by $p(x)$ in $F[x]$ is a maximal ideal of $F[x]$ if and only if $p(x)$ is irreducible in $F[x]$.

Proof. We first prove that if $p(x)$ is irreducible in $F[x]$, then the ideal $M = (p(x))$ is a maximal ideal of $F[x]$. For, suppose that N is an ideal of $F[x]$, and $N \supset M$. By Theorem 4.5.6,

$$N = (f(x)) \qquad \text{for some} \qquad f(x) \in F[x].$$

Because $p(x) \in M \subset N$, $p(x) = a(x)f(x)$, since every element in N is of this form. But $p(x)$ is irreducible in $F[x]$, hence $a(x)$ is a constant or $f(x)$ is a constant. If $a(x) = a \in F$, then $p(x) = af(x)$, so $f(x) = a^{-1}p(x)$, hence $f(x) \in M$, which says that $N \subset M$, hence $N = M$. On the other hand, if $f(x) = b \in F$, then $1 = b^{-1}b \in N$ is an ideal of $F[x]$,

thus $g(x)1 \in N$ for all $g(x) \in F[x]$. This says that $N = F[x]$. Therefore, we have shown M to be a maximal ideal of $F[x]$.

In the other direction, suppose that $M = (p(x))$ is a maximal ideal
of $F[x]$. If $p(x)$ is not irreducible, then $p(x) = a(x)b(x)$, where
$\deg a(x) \geq 1$, $\deg b(x) \geq 1$. Let $N = (a(x))$; then, since $p(x) =$
$a(x)b(x)$, $p(x) \in N$. Therefore, $M \subset N$. Since $\deg a(x) \geq 1$, $N =$
$(a(x)) \neq F[x]$, since every element in $(a(x))$ has degree at least that of
$a(x)$. By the maximality of M we conclude that $M = N$. But then
$a(x) \in N = M$, which tells us that $a(x) = f(x)p(x)$; combined with
$p(x) = a(x)b(x) = b(x)f(x)p(x)$, we get that $b(x)f(x) = 1$. Since
$\deg 1 = 0 < \deg b(x) \leq \deg(b(x)f(x)) = \deg 1 = 0$, we have reached a
contradiction. Thus $p(x)$ is irreducible. $\square$

This theorem is important because it tells us exactly what the maximal ideals of $F[x]$ are, namely the ideals generated by the irreducible
polynomials. If M is a maximal ideal of $F[x]$, $F[x]/M$ is a field, and
this field contains F (or more precisely, the field $\{a + M \mid a \in F\}$, which
is isomorphic to F). This allows us to construct decent fields $K \supset F$, the
decency of which lies in that $p(x)$ has a *root* in K. The exact statement
and explanation of this we postpone until Chapter 5.

The last topic in this direction that we want to discuss is the
factorization of a given polynomial as a product of irreducible ones.
Note that if $p(x) = a_0x^n + a_1x^{n-1} + \cdots + a_{n-1}x + a_n$, $a_0 \neq 0$, is
irreducible in $F[x]$, then so is $a_0^{-1}p(x)$ irreducible in $F[x]$; however,
$a_0^{-1}p(x)$ has the advantage of being monic. So we have this monic
irreducible polynomial trivially obtainable from $p(x)$ itself. This will
allow us to make more precise the uniqueness part of the next theorem.

Theorem 4.5.12. Let $f(x) \in F[x]$ be of positive degree.
Then either $f(x)$ is irreducible in $F[x]$ or $f(x)$ is the product
of irreducible polynomials in $F[x]$. In fact, then,

$$f(x) = ap_1(x)^{m_1}p_2(x)^{m_2} \cdots p_k(x)^{m_k},$$

where a is the highest coefficient of $f(x)$, $p_1(x), \ldots, p_k(x)$
are monic and irreducible in $F[x]$, $m_1 > 0, \ldots, m_k > 0$ and
this factorization in this form is unique up to the order of the
$p_i(x)$.

Proof. We first show the first half of the theorem, namely that $f(x)$ is irreducible or the product of irreducibles. The proof is exactly the same as that of Theorem 1.5.7, with a slight, obvious adjustment.

We go by induction on $\deg f(x)$. If $\deg f(x) = 1$, then $f(x) = ax + b$ with $a \neq 0$ and is clearly irreducible in $F[x]$. So the result is true in this case.

Suppose, then, that the theorem is correct for all $a(x) \in F[x]$ such that $\deg a(x) < \deg f(x)$. If $f(x)$ is irreducible, then we have nothing to prove. Otherwise, $f(x) = a(x)b(x)$, $a(x)$ and $b(x) \in F[x]$ and $\deg a(x) < \deg f(x)$ and $\deg b(x) < \deg f(x)$. By the induction, $a(x)$ [and $b(x)$] is irreducible or is the product of irreducibles. But then $f(x)$ is the product of irreducible polynomials in $F[x]$. This completes the induction, and so proves the opening half of the theorem.

Now to the uniqueness half. Again we go by induction on $\deg f(x)$. If $\deg f(x) = 1$, then $f(x)$ is irreducible and the uniqueness is clear.

Suppose the result true for polynomials of degree less than $\deg f(x)$. Suppose that

$$f(x) = ap_1(x)^{m_1} p_2(x)^{m_2} \cdots p_k(x)^{m_k} = aq_1(x)^{n_1} \cdots q_r(x)^{n_r},$$

where the $p_i(x)$, $q_i(x)$ are monic irreducible polynomials and the m_i, n_i are all positive and a is the highest coefficient of $f(x)$. Since $p_1(x)|f(x)$, we have that $p_1(x)|q_1(x)^{n_1} \cdots q_r(x)^{n_r}$, so by the corollary to Theorem 4.5.11, $p_1(x)|q_i(x)$ for some i. Since $q_i(x)$ is monic and irreducible, as is $p_1(x)$, we get $p_1(x) = q_i(x)$. We can suppose (on renumbering) that $p_1(x) = q_1(x)$. Thus

$$\frac{f(x)}{p_1(x)} = ap_1(x)^{m_1-1} p_2(x)^{m_2} \cdots p_k(x)^{m_k}$$

$$= ap_1(x)^{n_1-1} q_2(x)^{n_2} \cdots q_r(x)^{n_r}.$$

By induction we have unique factorization in the required form for $f(x)/p_1(x)$, whose degree is less than $\deg f(x)$. Hence we obtain that $m_1 - 1 = n_1 - 1$ (so $m_1 = n_1$), $m_2 = n_2, \ldots, m_k = n_k$, $r = k$ and $p_2(x) = q_2(x), \ldots, p_k(x) = q_k(x)$, on renumbering the q's appropriately. This completes the induction and proves the theorem. $\square$

We have pointed out how similar the situation is for the integers $\mathbb{Z}$ and the polynomial ring $F[x]$. This suggests that there should be a wider class of rings, of which the two examples $\mathbb{Z}$ and $F[x]$ are special cases, for which much of the argumentation works. It worked for Z and $F[x]$ because we had a measure of size in them, either by the size of an integer or the degree of a polynomial. This measure of size was such that it allowed a Euclid-type algorithm to hold.

This leads us to define a class of rings, the *Euclidean rings*.

> **Definition.** An integral domain R is a *Euclidean ring* if there is a function d from the *nonzero* elements of R to the nonnegative integers that satisfies:
> (a) For $a \neq 0$, $b \neq 0 \in R$, $d(a) \leq d(ab)$.
> (b) Given $a \neq 0$, $b \neq 0$, there exist q and $r \in R$ such that $b = qa + r$, where $r = 0$ or $d(r) < d(a)$.

The interested student should try to see which of the results proved for polynomial rings (and the integers) hold in a general Euclidean ring. Aside from a few problems involving Euclidean rings, we shall not go any further with this interesting class of rings.

The final comment we make here is that what we did for polynomials over a field we could *try* to do for polynomials over an arbitrary ring. That is, given R any ring (commutative or noncommutative), we could define the polynomial ring $R[x]$ in x over R by defining equality, addition, and multiplication exactly as we did in $F[x]$, for F a field. The ring so constructed, $R[x]$, is a very interesting ring, whose structure is tightly interwoven with that of R itself. It would be too much to expect that all, or even any, of the theorems proved in this section would carry over to $R[x]$ for a general ring R.

Problems

In the following problems, F will always denote a field.

Easier Problems

1. If F is a field, show that the only invertible elements in $F[x]$ are the nonzero elements of F.

2. If R is a ring, we introduce the ring $R[x]$ of polynomials in x over R, just as we did $F[x]$. Defining $\deg f(x)$ for $f(x) \in R[x]$ as we did in $F[x]$, show that:
   - (a) $\deg(f(x)g(x)) \le \deg f(x) + \deg g(x)$ if $f(x)g(x) \ne 0$.
   - (b) There is a commutative ring R such that we can find $f(x)$, $g(x)$ in $R[x]$ with $\deg(f(x)g(x)) < \deg f(x) + \deg g(x)$.

3. Find the greatest common divisor of the following polynomials over $\mathbb{Q}$, the field of rational numbers.
   - (a) $x^3 - 6x + 7$ and $x + 4$.
   - (b) $x^2 - 1$ and $2x^7 - 4x^5 + 2$.
   - (c) $3x^2 + 1$ and $x^6 + x^4 + x + 1$.
   - (d) $x^3 - 1$ and $x^7 - x^4 + x^3 - 1$.

4. Prove Lemma 4.5.3.

5. In Problem 3, let $I = \{f(x)a(x) + g(x)b(x)\}$, where $f(x)$, $g(x)$ run over $\mathbb{Q}[x]$ and $a(x)$ is the first polynomial and $b(x)$ the second one in each part of the problem. Find $d(x)$, so that $I = (d(x))$ for Parts (a), (b), (c), and (d).

6. If $g(x)$, $f(x) \in F[x]$ and $g(x)|f(x)$, show that $(f(x)) \subset (g(x))$.

7. Prove the uniqueness of the greatest common divisor of two polynomials in $F[x]$ by using Lemma 4.5.8.

8. If $f(x)$, $g(x) \in F[x]$ are relative prime and $f(x)|h(x)$ and $g(x)|h(x)$, show that $f(x)g(x)|h(x)$.

9. Prove the Corollary to Theorem 4.5.10.

10. Show that the following polynomials are irreducible over the field F indicated.
   - (a) $x^2 + 7$ over $F =$ real field $= \mathbb{R}$.
   - (b) $x^3 - 3x + 3$ over $F =$ rational field $= \mathbb{Q}$.
   - (c) $x^2 + x + 1$ over $F = \mathbb{Z}_2$.
   - (d) $x^2 + 1$ over $F = \mathbb{Z}_{19}$.
   - (e) $x^3 - 9$ over $F = \mathbb{Z}_{13}$.
   - (f) $x^4 + 2x^2 + 2$ over $F = \mathbb{Q}$.

11. If $p(x) \in F[x]$ is of degree 3 and $p(x) = a_0x^3 + a_1x^2 + a_2x + a_3$, show that $p(x)$ is irreducible over F if there is no element $r \in F$ such that $p(r) = a_0r^3 + a_1r^2 + a_2r + a_3 = 0$.

12. If $F \subset K$ are two fields and $f(x), g(x) \in F[x]$ are relatively prime in $F[x]$, show that they are relatively prime in $K[x]$.

Middle-Level Problems

13. Let $\mathbb{R}$ be the field of real numbers and $\mathbb{C}$ that of complex numbers. Show that $\mathbb{R}[x]/(x^2 + 1) \simeq \mathbb{C}$. [**Hint:** If $A = \mathbb{R}[x]/(x^2 + 1)$, let u be the image of x in A; show that every element in A is of the form $a + bu$, where $a, b \in \mathbb{R}$ and $u^2 = -1$.]

14. Let $F = \mathbb{Z}_{11}$, the integers mod 11.
    (a) Let $p(x) = x^2 + 1$; show that $p(x)$ is irreducible in $F[x]$ and that $F[x]/(p(x))$ is a field having 121 elements.
    (b) Let $p(x) = x^3 + x + 4 \in F[x]$; show that $p(x)$ is irreducible in $F[x]$ and that $F[x]/(p(x))$ is a field having 11^3 elements.

15. Let $F = \mathbb{Z}_p$ be the field of integers mod p, where p is a prime, and let $q(x) \in F[x]$ be irreducible of degree n. Show that $F[x]/(q(x))$ is a field having at most p^n elements. (See Problem 16 for a more exact statement.)

16. Let F, $q(x)$ be as in Problem 15; shows that $F[x]/(q(x))$ has exactly p^n elements.

17. Let $p_1(x), p_2(x), \ldots, p_k(x) \in F[x]$ be distinct irreducible polynomials and let $q(x) = p_1(x)p_2(x) \cdots p_k(x)$. Show that

$$\frac{F[x]}{(q(x))} \simeq \frac{F[x]}{(p_1(x))} \oplus \frac{F[x]}{(p_2(x))} \oplus \cdots \oplus \frac{F[x]}{(p_k(x))}.$$

18. Let F be a finite field. Show that $F[x]$ contains irreducible polynomials of arbitrarily high degree. (**Hint:** Try to imitate Euclid's proof that there is an infinity of prime numbers.)

19. Construct a field having p^2 elements, for p an odd prime.

20. If R is a Euclidean ring, show that every ideal of R is principal.

21. If R is a Euclidean ring, show that R has a unit element.

22. If R is the ring of even integers, show that Euclid's algorithm is false in R by exhibiting two even integers for which the algorithm does not hold.

Harder Problems

23. Let $F = \mathbb{Z}_7$ and let $p(x) = x^3 - 2$ and $q(x) = x^3 + 2$ be in $F[x]$. Show that $p(x)$ and $q(x)$ are irreducible in $F[x]$ and that the fields $F[x]/(p(x))$ and $F[x]/(q(x))$ are isomorphic.

24. Let $\mathbb{Q}$ be the field of rational numbers, and let $q(x) = x^2 + x + 1$ in $\mathbb{Q}(x)$. If α is a complex number such that $\alpha^2 + \alpha + 1 = 0$, show that the set $\{a + b\alpha |\, a, b \in \mathbb{Q}\}$ is a field in two ways; the first by showing it to be isomorphic to something you know is a field, the second by showing that if $a + b\alpha \neq 0$, then its inverse is of the same form.

25. If p is a prime, show that $q(x) = 1 + x + x^2 + \cdots x^{p-1}$ is irreducible in $\mathbb{Q}[x]$.

26. Let R be a commutative ring in which $a^2 = 0$ only if $a = 0$. Show that if $q(x) \in R[x]$ is a zero-divisor in $R[x]$, then, if

$$q(x) = a_0 x^n + a_1 x^{n-1} + \cdots + a_n,$$

there is an element $b \neq 0$ in R such that $ba_0 = ba_1 = \cdots ba_n = 0$.

27. Let R be a ring and I an ideal of R. If $R[x]$ and $I[x]$ are the polynomial rings in x over R and I, respectively, show that:
(a) $I[x]$ is an ideal of $R[x]$.
(b) $R[x]/I[x] \simeq (R/I)[x]$.

Very Hard Problems

*28. Do Problem 26 even if the condition "$a^2 = 0$ only if $a = 0$" does not hold in R.

29. Let $R = \{a + bi |\, a, b$ integers$\} \subset \mathbb{C}$. Let $d(a + bi) = a^2 + b^2$. Show that R is a Euclidean ring with this d its required Euclidean function. (R is known as the ring of *Gaussian integers* and plays an important role in number theory.)

6.   POLYNOMIALS OVER THE RATIONALS

In our consideration of the polynomial ring $F[x]$ over a field F, the particular nature of F never entered the picture. All the results hold for arbitrary fields. However, there are results that exploit the explicit character of certain fields. One such field is that of the rational numbers.

We shall present two important theorems for $\mathbb{Q}[x]$, the polynomial ring over the rational field $\mathbb{Q}$. These results depend heavily on the fact that we are dealing with rational numbers. The first of these, *Gauss' Lemma*, relates the factorization over the rationals with factorization over the integers. The second one, known as the *Eisenstein Criterion*, gives us a method of constructing irreducible polynomials of arbitrary degree, at will, in $\mathbb{Q}[x]$. In this the field $\mathbb{Q}$ is highly particular. For instance, there is no easy algorithm for obtaining irreducible polynomials of arbitrary degree n over the field $\mathbb{Z}_p$ of the integers mod p, p a prime. Even over $\mathbb{Z}_2$ such an algorithm is nonexistent; it would be highly useful to have, especially for coding theory. But it just doesn't exist—so far.

We begin our consideration with two easy results.

Lemma 4.6.1.   Let $f(x) \in \mathbb{Q}[x]$; then

$$f(x) = \frac{u}{m}\left(a_0 x^n + a_1 x^{n-1} + \cdots + a_n\right)$$

where $u, m, a_0, \ldots, a_n$ are integers and the $a_0, a_1, \ldots, a_n$ have no common factor greater than 1 (i.e., are relatively prime) and $(u, m) = 1$.

Proof.   Since $f(x) \in \mathbb{Q}[x]$, $f(x) = q_0 x^n + q_1 x^{n-1} + \cdots + q_n$, where the q_i are rational numbers. So for $i = 0, 1, 2, \ldots, n$, $q_i = b_i/c_i$, where b_i, c_i are integers. Thus

$$f(x) = \frac{b_0}{c_0} x^n + \frac{b_1}{c_1} x^{n-1} + \cdots + \frac{b_n}{c_n};$$

clearing of denominators gives us

$$f(x) = \frac{1}{c_0 c_1 \cdots c_n}\left(u_0 x^n + u_1 x^{n-1} + \cdots + u_n\right),$$

where the u_i are integers. If w is the greatest common divisor of $u_0, u_1, \ldots, u_n$, then each $u_i = w a_i$, where $a_0, a_1, \ldots, a_n$ are relatively prime integers. Then

$$f(x) = \frac{w}{c_0 c_1 c_2 \cdots c_n} (a_0 x^n + a_1 x^{n-1} + \cdots + a_n);$$

canceling out the greatest common factor of w and $c_0 c_1 \cdots c_n$ gives us

$$f(x) = \frac{u}{m} (a_0 x^n + \cdots + a_n),$$

where u, m are relatively prime integers, as is claimed in the lemma. $\square$

The next result is a result about a particular homomorphic image of $R[x]$ for any ring R.

Lemma 4.6.2. If R is any ring and I an ideal of R, then $I[x]$, the polynomial ring in x over I, is an ideal of $R[x]$. Furthermore, $R[x]/I[x] \simeq (R/I)[x]$, the polynomial ring in x over R/I.

Proof. Let $\overline{R} = R/I$; then there is a homomorphism $\varphi: R \to \overline{R}$, defined by $\varphi(a) = a + I$, whose kernel is I. Define $\Phi: R[x] \to \overline{R}[x]$ by: If

$$f(x) = a_0 x^n + a_1 x^{n-1} + \cdots a_n,$$

then

$$\Phi(f(x)) = \varphi(a_0) x^n + \varphi(a_1) x^{n-1} + \cdots + \varphi(a_n).$$

We leave it to the reader to prove that Φ is a homomorphism of $R[x]$ onto $\overline{R}[x]$. What is the kernel, $K(\Phi)$, of Φ? If $f(x) = a_0 x^n + \cdots + a_n$ is in $K(\Phi)$, then $\Phi(f(x)) = 0$, the 0 element of $\overline{R}[x]$. Since

$$\Phi(f(x)) = \varphi(a_0) x^n + \varphi(a_1) x^{n-1} + \cdots + \varphi(a_n) = 0,$$

each coefficient $\varphi(a_0) = 0$, $\varphi(a_1) = 0, \ldots, \varphi(a_n) = 0$, by the very definition of what we mean by the 0-polynomial in a polynomial ring. Thus each a_i is in the kernel of φ, which happens to be I. Because

$a_0, a_1, \ldots, a_n$ are in I, $f(x) = a_0 x^n + a_1 x^{n-1} + \cdots + a_n$ is in $I[x]$. So $K(\Phi) \subset I[x]$. That $I[x] \subset K(\Phi)$ is immediate from the definition of the mapping Φ. Hence $I[x] = K(\Phi)$. By the First Homomorphism Theorem (Theorem 4.3.3), the ring $I[x]$ is then an ideal of $R[x]$ and $\overline{R}[x] \simeq R[x]/K(\Phi) = R[x]/I[x]$. This proves the lemma, remembering that $\overline{R} = R/I$. $\square$

As a very special case of the lemma we have the

Corollary. Let $\mathbb{Z}$ be the ring of integers, p a prime number in $\mathbb{Z}$, and $I = (p)$, the ideal of $\mathbb{Z}$ generated by p. Then $\mathbb{Z}[x]/I[x] \simeq \mathbb{Z}_p[x]$.

Proof. Since $\mathbb{Z}_p \simeq \mathbb{Z}/I$, the corollary follows by applying the lemma to $R = \mathbb{Z}$. $\square$

We are ready to prove the first of the two major results we seek in this section.

Theorem 4.6.3 (Gauss' Lemma). If $f(x) \in \mathbb{Z}[x]$ is a monic polynomial and $f(x) = a(x)b(x)$, where $a(x)$ and $b(x)$ are in $\mathbb{Q}[x]$, then $f(x) = a_1(x)b_1(x)$, where $a_1(x), b_1(x)$ are monic polynomials in $\mathbb{Z}[x]$ and $\deg a_1(x) = \deg a(x)$, $\deg b_1(x) = \deg b(x)$.

Proof. Suppose that $f(x) = x^n + u_1 x^{n-1} + \cdots + u_n$, where the $u_i \in \mathbb{Z}$ are integers. Because $a(x)$ and $b(x)$ are in $\mathbb{Q}[x]$, $a(x) = a_0 x^s + a_1 x^{s-1} + \cdots + a_s$ and $b(x) = b_0 x^r + b_1 x^{r-1} + \cdots + b_r$, where the a_i's and b_j's are rational numbers. By Lemma 4.6.1,

$$a(x) = \frac{u_1}{m_1}\left(a_0' x^s + a_1' x^{s-1} + \cdots + a_s'\right) = \frac{u_1}{m_1} a_1(x),$$

where $a_0', a_1', \ldots, a_s'$ are relatively prime integers and

$$b(x) = \frac{u_2}{m_2}\left(b_0' x^r + b_1' x^{r-1} + \cdots + b_r'\right) = \frac{u_2}{m_2} b_1(x),$$

where $b'_0, b'_1, \ldots, b'_r$ are relatively prime. Thus

$$f(x) = a(x)b(x) = \frac{u_1 u_2}{m_1 m_2} a_1(x) b_1(x) = \frac{v}{w} a_1(x) b_1(x),$$

where v and w are relatively prime, by canceling out the common factor of $u_1 u_2$ and $m_1 m_2$. Therefore, $wf(x) = va_1(x)b_1(x)$, and $f(x), a_1(x), b_1(x)$ are all in $\mathbb{Z}[x]$.

If $w = 1$, then, since $f(x)$ is monic, we get that $va'_0 b'_0 = 1$ and this leads easily to $v = 1$, $a'_0 = b'_0 = 1^\dagger$ and so $f(x) = a_1(x)b_1(x)$, where both $a_1(x)$ and $b_1(x)$ are monic polynomials with integer coefficients. This is precisely the claim of the theorem, since $\deg a_1(x) = \deg a(x)$ and $\deg b_1(x) = \deg b(x)$.

Suppose then that $w \neq 1$; thus there is a prime p such that $p|w$ and, since $(v, w) = 1$, $p \nmid v$. Also, since the coefficients $a'_0, a'_1, \ldots, a'_s$ of $a_1(x)$ are relatively prime, there is an i such that $p \nmid a'_i$; similarly, there is a j such that $p \nmid b'_j$. Let $I = (p)$ be the ideal generated by p in $\mathbb{Z}$; then $\mathbb{Z}/I \simeq \mathbb{Z}_p$ and, by the Corollary to Lemma 4.6.2, $\mathbb{Z}[x]/I[x] \simeq \mathbb{Z}_p[x]$, so is an integral domain. However, since $p|w$, $\bar{w}$, the image of w in $\mathbb{Z}[x]/I[x]$, is 0, and since $p \nmid v$, $\bar{v}$ the image of v in $\mathbb{Z}[x]/I[x]$ is not 0. Thus $0\bar{f}(x) = \bar{v}\bar{a}_1(x)\bar{b}_1(x)$, where $\bar{v} \neq 0$ and $\bar{a}_1(x) \neq 0$, $\bar{b}_1(x) \neq 0$ because $p \nmid a'_i$ and $p \nmid b'_j$ for the given i, j above. This contradicts that $\mathbb{Z}[x]/I[x]$ is an integral domain. So $w \neq 1$ is not possible, and the theorem is proved. $\square$

It might be instructive for the reader to try to show directly that if $x^3 + 6x - 7$ is the product of two polynomials having rational coefficients, then it is already the product of two monic polynomials with integer coefficients.

> *One should say something about C. F. Gauss (1777–1855), considered by many the greatest mathematician ever. His contributions in number theory, algebra, geometry, and so on, are of gigantic proportions. His contributions in physics and astronomy are also of such a great proportion that he is considered by physicists as one of their greats, and by the astronomers as one of the important astronomers of the past.*

As we indicated at the beginning of this section, irreducible polynomials of degree n over a given field F may be very hard to come by.

However, over the rationals, due to the next theorem, these exist in abundance and are very easy to construct.

---

† $a_0' = b_0' = -1$ is also possible, in which case $-a_1(x)$, $-b_1(x)$ are monic polynomials and $f(x) = (-a_1(x))(-b_1(x))$.

Theorem 4.6.4 (The Eisenstein Criterion). Let $f(x) = x^n + a_1x^{n-1} + \cdots + a_n$ be a polynomial with integer coefficients. Suppose that there is some prime p such that $p|a_1, p|a_2, \ldots, p|a_n$, but $p^2 \nmid a_n$. Then $f(x)$ is irreducible in $\mathbb{Q}[x]$.

Proof. Suppose that $f(x) = u(x)v(x)$, where $u(x)$, $v(x)$ _are of positive degree_ and are polynomials in $\mathbb{Q}[x]$. By Gauss' Lemma we may assume that both $u(x)$ and $v(x)$ are monic polynomials with integer coefficients. Let $I = (p)$ be the ideal generated by p in Z, and consider $\mathbb{Z}[x]/I[x]$, which is an integral domain, since we know by the Corollary to Lemma 4.6.2 that $\mathbb{Z}[x]/I[x] \simeq (\mathbb{Z}/I)[x] \simeq \mathbb{Z}_p[x]$. The image of $f(x) = x^n + a_1x^{n-1} + \cdots + a_n$ in $Z[x]/I[x]$ is x^n, since $p|a_1, \ldots, p|a_n$. So if $\bar{u}(x)$ is the image of $u(x)$ and $\bar{v}(x)$ that of $v(x)$ in $\mathbb{Z}[x]/I[x]$, then $x^n = \bar{u}(x)\bar{v}(x)$. Since $\bar{u}(x)|x^n$, $\bar{v}(x)|x^n$ in $\mathbb{Z}[x]/I[x]$, we must have that $\bar{u}(x) = x^r$, $\bar{v}(x) = x^{n-r}$ for some $1 < r < n$. But then $u(x) = x^r + pg(x)$ and $v(x) = x^{n-r} + ph(x)$, where $g(x)$ and $h(x)$ are polynomials with integer coefficients. Since $u(x)v(x) = x^n + px^rh(x) + px^{n-r}g(x) + p^2g(x)h(x)$, and since $1 < r < n$, the constant term of $u(x)v(x)$ is p^2st, where s is the constant term of $g(x)$ and t the constant term of $h(x)$. Because $f(x) = u(x)v(x)$, their constant terms are equal, hence $a_n = p^2st$. Since s and t are integers, we get that $p^2|a_n$, a contradiction. In this way we see that $f(x)$ is irreducible. $\square$

We give some examples of the use to which the Eisenstein Criterion can be put.

1. Let $f(x) = x^n - p$, p any prime. Then one sees at a glance that $f(x)$ is irreducible in $\mathbb{Q}[x]$, for the Eisenstein Criterion applies.

2. Let $f(x) = x^5 - 4x + 22$. Since $2|22$, $2^2 \nmid 22$ and 2 divides the other relevant coefficients of $f(x)$, the Eisenstein Criterion tells us that $f(x)$ is irreducible in $\mathbb{Q}[x]$.

3. Let $f(x) = x^{11} - 6x^4 + 12x^3 + 36 - 6$. We see that $f(x)$ is irreducible in $\mathbb{Q}[x]$ by using either 2 or 3 to check the conditions of the Eisenstein Criterion.

4. Let $f(x) = 5x^4 - 7x + 7$; $f(x)$ is not monic, but we can modify $f(x)$ slightly to be in a position where we can apply the Eisenstein Criterion. Let

$$g(x) = 5^3 f(x) = 5^4 x^4 - 7 \cdot 5^3 x + 7 \cdot 5^3 = (5x)^4 - 175(5x) + 875;$$

if we let $y = 5x$, then $g(x) = h(y) = y^4 - 175y + 875$. The polynomial $h(y)$ is irreducible in $\mathbb{Z}[y]$ by using the prime 7 and applying the Eisenstein Criterion. The irreducibility of $h(y)$ implies that of $g(x)$, and so that of $f(x)$, in $\mathbb{Q}[x]$.

This suggests a slight generalization of the Eisenstein Criterion to nonmonic polynomials. (See Problem 4.)

5. Let $f(x) = x^4 + x^3 + x^2 + x + 1$; as it stands we cannot, of course, apply the Eisenstein Criterion to $f(x)$. We pass to a polynomial $g(x)$ closely related to $f(x)$ whose irreducibility in $\mathbb{Q}[x]$ will ensure that of $f(x)$. Let $g(x) = f(x + 1) = (x + 1)^4 + (x + 1)^3 + (x + 1)^2 + (x + 1) + 1 = x^4 + 5x^3 + 10x^2 + 10x + 5$. The Eisenstein Criterion applies to $g(x)$, using the prime 5; thus $g(x)$ is irreducible in $\mathbb{Q}[x]$. This implies that $f(x)$ is irreducible in $\mathbb{Q}[x]$. (See Problem 1.)

Gotthold Eisenstein (1823–1852) in his short life made fundamental contributions in algebra and analysis.

Problems

1. In Example, 5, show that because $g(x)$ is irreducible in $\mathbb{Q}[x]$, then so is $f(x)$.

2. Prove that $f(x) = x^3 + 3x + 2$ is irreducible in $\mathbb{Q}[x]$.

3. Show that there is an infinite number of integers a such that $f(x) = x^7 + 15x^2 - 30x + a$ is irreducible in $\mathbb{Q}[x]$. What a's do you suggest?

4. Prove the following generalization of the Eisenstein Criterion. Let $f(x) = a_0 x^n + a_1 x^{n-1} + \cdots + a_n$ have integer coefficients and suppose that there is a prime p such that

$$p \nmid a_0, \ p | a_1, \ p | a_2, \ldots, p | a_{n-1}, \ p | a_n,$$

but $p^2 \nmid a_n$; then $f(x)$ is irreducible in $\mathbb{Q}[x]$.

5. If p is a prime, show that $f(x) = x^{p-1} + x^{p-2} + \cdots + x + 1$ is irreducible in $Q[x]$.

6. Let F be a field and φ an automorphism of $F[x]$ such that $\varphi(a) = a$ for every $a \in F$. If $f(x) \in F[x]$, prove that $f(x)$ is irreducible in $F[x]$ if and only if $g(x) = \varphi(f(x))$ is.

7. Let F be a field. Define the mapping

$$\varphi: F[x] \to F[x] \text{ by } \varphi(f(x)) = f(x + 1)$$

for every $f(x) \in F[x]$. Prove that φ is an automorphism of $F[x]$ such that $\varphi(a) = a$ for every $a \in F$.

8. Let F be a field and $b \neq 0$ an element of F. Define the mapping $\varphi: F[x] \to F[x]$ by $\varphi(f(x)) = f(bx)$ for every $f(x) \in F[x]$. Prove that φ is an automorphism of $F[x]$ such that $\varphi(a) = a$ for every $a \in F$.

9. Let F be a field, $b \neq 0, c$ elements of F. Define the mapping $\varphi: F[x] \to F[x]$ by $\varphi(f(x)) = f(bx + c)$ for every $f(x) \in F[x]$. Prove that φ is an automorphism of $F[x]$ such that $\varphi(a) = a$ for every $a \in F$.

10. Let φ be an automorphism of $F[x]$, where F is a field, such that $\varphi(a) = a$ for every $a \in F$. Prove that if $f(x) \in F[x]$, then $\deg \varphi(f(x)) = \deg f(x)$.

11. Let φ be an automorphism of $F[x]$, where F is a field, such that $\varphi(a) = a$ for every $a \in F$. Prove there exist $b \neq 0, c$ in F such that $\varphi(f(x)) = f(bx + c)$ for every $f(x) \in F[x]$.

12. Find a nonidentity automorphism φ of $\mathbf{Q}[x]$ such that φ^2 is the identity automorphism, of $\mathbf{Q}[x]$.

13. Show that in Problem 12 you do not need the assumption $\varphi(a) = a$ for every $a \in \mathbf{Q}$ because any automorphism of $\mathbf{Q}[x]$ automatically satisfies $\varphi(a) = a$ for every $a \in \mathbf{Q}$.

14. Let $\mathbf{C}$ be the field of complex numbers. Given an integer $n > 0$, exhibit an automorphism φ of $\mathbf{C}[x]$ of order n.

7.   FIELD OF QUOTIENTS OF AN INTEGRAL DOMAIN

Given the integral domain Z, the ring of integers, then intimately related to Z is the field Q of rational numbers that consists of all fractions of integers; that is, all quotients m/n, where $m, n \neq 0$ are in Z. Note that there is no unique way of representing $\frac{1}{2}$, say, in Q because $\frac{1}{2} = \frac{2}{4} = (-7)/(-14) = \cdots$. In other words, we are identifying $\frac{1}{2}$ with $\frac{2}{4}, (-7)/(-14)$, and so on. This suggests that what is really going on in constructing the rationals from the integers is some equivalence relation on some set based on the integers.

The relation of Q to Z carries over to any integral domain D. Given an integral domain D, we shall construct a field $F \supset D$ whose elements will be quotients a/b with $a, b \neq 0 \in D$. We go through this construction formally.

Let D be an integral domain, and let $S = \{(a, b)|a, b \in D, b \neq 0\}$; S is thus the subset of $D \times D$—the Cartesian product of D with itself —in which the second component is not allowed to be 0. Think of (a, b) as a/b for a moment; if so, when would we want to declare that $(a, b) = (c, d)$? Clearly, we would want this if $a/b = c/d$, which in D itself would become $ad = bc$. With this as our motivating guide we define a relation $\sim$ on S by declaring:

$$(a, b) \sim (c, d) \text{ for } (a, b), (c, d) \text{ in } S \text{ if and only if } ad = bc.$$

We first assert that this defines an equivalence relation on S. We go through the three requirements for an equivalence relation term by term.

1. $(a, b) \sim (a, b)$, for clearly $ab = ba$ (since D is commutative). So $\sim$ is reflexive.

2. $(a, b) \sim (c, d)$ implies that $(c, d) \sim (a, b)$, for $(a, b) \sim (c, d)$ means $ad = bc$; for $(c, d) \sim (a, b)$ we need $cb = da$, but this is true, since $cb = bc = ad = da$. So $\sim$ is symmetric.

3. $(a, b) \sim (c, d)$, $(c, d) \sim (e, f)$ implies that $ad = bc$, $cf = de$, so $adf = bcf = bde$; but $d \neq 0$ and we are in an integral domain, hence $af = be$ follows. This says that $(a, b) \sim (e, f)$. So the relation is transitive.

We have shown that $\sim$ defines an equivalence relation on S. Let F be the set of all the equivalence classes $[a, b]$ of the elements $(a, b) \in S$. F is our required field.

To show that F is a field, we must endow it with an addition and multiplication. First the addition; what should it be? Forgetting all the fancy talk about equivalence relation and the like, we really want $[a, b]$ to be a/b. If so, what should $[a, b] + [c, d]$ be other than the formally calculated

$$\frac{a}{b} + \frac{c}{d} = \frac{ad + bc}{bd}.$$

This motivates us to *define*

(1) $$[a, b] + [c, d] = [ad + bc, bd].$$

Note that since $b \neq 0$, $d \neq 0$ and D is a domain, then $bd \neq 0$, hence $[ad + bc, bd]$ is a legitimate element of F.

As usual we are plagued with having to show that the addition so defined in F is well-defined. In other words, we must show that if $[a, b] = [a', b']$ and $[c, d] = [c', d']$, then $[a, b] + [c, d] = [a', b'] + [c', d']$. From (1) we must thus show that $[ad + bc, bd] = [a'd' + b'c', b'd']$, which is to say, $(ad + bc)b'd' = bd(a'd' + b'c')$. Since $[a, b] = [a', b']$ and $[c, d] = [c', d']$, $ab' = ba'$ and $cd' = dc'$. Therefore, $(ad + bc)b'd' = ab'dd' + bb'cd' = ba'dd' + bb'dc' = (a'd' + b'c')bd$, as required. Thus "+" is well-defined in F.

The class $[0, b]$, $b \neq 0$, acts as the 0 under "+," we denote it simply as 0, and the class $[-a, b]$ is the negative of $[a, b]$. To see that this makes of F an abelian group is easy, but laborious, for all that really needs verification is the associative law.

Now to the multiplication in F. Again motivated by thinking of $[a, b]$ as a/b, we *define*

(2) $$[a, b][c, d] = [ac, bd].$$

Again since $b \neq 0$, $d \neq 0$, we have $bd \neq 0$, so the element $[ac, bd]$ is also a legitimate element of F.

As for the "+" we must show that the product so introduced is well-defined; that is, if $[a, b] = [a', b']$, $[c, d] = [c', d']$, then

$$[a, b][c, d] = [ac, bd] = [a'c', b'd'] = [a', b'][c', d'].$$

We know that $ab' = ba'$ and $cd' = dc'$, so $acb'd' = ab'cd' = ba'dc' = bda'c'$, which is exactly what we need for $[ac, bd] = [a'c', b'd']$. Thus the product is well-defined in F.

What acts as 1 in F? We claim that for any $a \neq 0$, $b \neq 0$ in D, $[a, a] = [b, b]$ (since $ab = ba$) and $[c, d][a, a] = [ca, da] = [c, d]$, since $(ca)d = (da)c$. So $[a, a]$ acts as 1, and we write it simply as $1 = [a, a]$ (for all $a \neq 0$ in D). Given $[a, b] \neq 0$, then $a \neq 0$, so $[b, a]$ is in F; hence, because $[a, b][b, a] = [ab, ba] = [ab, ab] = 1$, $[a, b]$ has an inverse in F. All that remains to show that the nonzero elements of F form an abelian group under this product is the associative law and commutative law. We leave these to the reader.

To clinch that F is a field, we need only now show the distributive law. But $[ad + bc, bd][e, f] = [(ad + bc)e, bdf]$, so

$$([a, b] + [c, d])[e, f] = [ade + bce, bdf],$$

while

$$[a, b][e, f] + [c, d][e, f]$$

$$= [ae, bf] + [ce, df] = [aedf + bfce, bdf^2]$$

$$= [(ade + bce)f, bdf^2] = [ade + bcf, bdf][f, f]$$

$$= [ade + bce, bdf],$$

which we have seen is $([a, b] + [c, d])[e, f]$. The distributive law is now established, so F is a field.

Let $a \neq 0$ be a fixed element in D and consider $[da, a]$ for any $d \in D$. The map $\varphi : d \to [da, a]$ is a monomorphism of D into F. It is certainly 1-1, for if $\varphi(d) = [da, a] = 0$, then $da = 0$, so $d = 0$, since D is an integral domain. Also, $\varphi(d_1 d_2) = [d_1 d_2 a, a]$ while $\varphi(d_1)\varphi(d_2) = [d_1 a, a][d_2 a, a] = [d_1 d_2 a^2, a^2] = [d_1 d_2 a, a][a, a] = [d_1 d_2 a, a] = \varphi(d_1 d_2)$. Furthermore,

$$([d_1 a, a] + [d_2 a, a]) = ([d_1 a^2 + a^2 d_2, a^2])$$

$$= [d_1 a + d_2 a, a][a, a]$$

$$= ([(d_1 + d_2)a, a])$$

so $\varphi(d_1 + d_2) = [(d_1 + d_1)a, a] = [d_1 a, a] + [d_2 a, a] = \varphi(d_1) + \varphi(d_2)$. Thus φ maps D monomorphically into F. So, D is isomorphic to a subring of F, and we can thus consider D as "embedded" in F. We consider every element $[a, b]$ of F as the fraction a / b.

What we have shown is the

Theorem 4.7.1.   Let D be an integral domain. Then there exists a field $F \supset D$ which consists of all fractions a/b, as defined above, of elements in D.

The field F is called the *field of quotients* of D. When $D = Z$, then F is isomorphic to the field Q of rational numbers. Also, if D is the domain of even integers, then F is also the entire field Q.

What we did above in constructing the field of quotients of D was a long, formal, wordy, and probably dull way of doing something that is by its nature something very simple. We really are doing nothing more than forming all formal fractions a/b, $a, b \neq 0$ in D, where we add and multiply fractions as usual. However, it is sometimes necessary to see something done to its last detail, painful though it may be. Most of us had never seen a really formal and precise construction of the rationals from the integers. Now that we have constructed F from D in this formal manner, forget the details and think of F as the set of all fractions of elements of D.

Problems

1. Prove the associative law of addition in F.

2. Prove the commutative law of addition in F.

3. Prove that the product in F is commutative and associative.

4. If K is any field that contains D, show that $K \supset F$. (So F is the smallest field that contains D.)

CHAPTER 5

Fields

The notion of a ring was unfamiliar territory for most of us; that of a field touches more closely on our experience. While the only ring, other than a field, that we might have seen in our early training was the ring of integers, we had a bit more experience working with rational numbers, real numbers, and, some of us, complex numbers, in solving linear and quadratic equations. The ability to divide by nonzero elements gave us a bit of leeway, which we might not have had with the integers, in solving a variety of problems.

So at first glance, when we start working with fields we feel that we are on home ground. As we penetrate deeper into the subject, we start running across new ideas and new areas of results. Once again we'll be in unfamiliar territory, but hopefully, after some exposure to the topic, the notions will become natural to us.

Fields play an important role in geometry, in the theory of equations, and in certain very important parts of number theory. We shall touch upon each of these aspects as we progress. Unfortunately, because of the technical machinery we would need to develop, we do not go into Galois theory, a very beautiful part of the subject. We hope that many of the readers will make contact with Galois theory, and beyond, in their subsequent mathematical training.

1. EXAMPLES OF FIELDS

Let's recall that a field F is a commutative ring with unit element 1 such that for every nonzero $a \in F$ there is an element $a^{-1} \in F$ such that $aa^{-1} = 1$. In other words, fields are "something like" the rationals $\mathbb{Q}$. But are they really? The integers mod p, Z_p, where p is a prime, form a field; in Z_p we have the relation

$$p1 = \underbrace{1 + 1 + \cdots + 1}_{(p \text{ times})} = 0.$$

Nothing akin to this happens in Q. There are even sharper differences among fields—how polynomials factor over them, special properties of which we'll see some examples, and so on.

We begin with some familiar examples.

EXAMPLES

1. $\mathbb{Q}$, the field of rational numbers.

2. $\mathbb{R}$, the field of real numbers.

3. $\mathbb{C}$, the field of complex numbers.

4. Let $F = \{a + bi | a, b \in Q\} \subset \mathbb{C}$. To see that F is a field is relatively easy. We only verify that if $a + bi \neq 0$ is in F, then $(a + bi)^{-1}$ is also in F. But what is $(a + bi)^{-1}$? It is merely

$$\frac{a}{(a^2 + b^2)} - \frac{ib}{(a^2 + b^2)} \qquad \text{(verify!)},$$

and since $a^2 + b^2 \neq 0$ and is rational, therefore $a/(a^2 + b^2)$ and also $b/(a^2 + b^2)$ are rational, hence $(a + bi)^{-1}$ is indeed in F.

5. Let $F = \{a + b\sqrt{2} | a, b \in Q\} \subset \mathbb{R}$. Again the verification that F is a field is not too hard. Here, too, we only show the existence of inverses in F for the nonzero elements of F. Suppose that $a + b\sqrt{2} \neq 0$ is in F;

then, *since $\sqrt{2}$ is irrational,* $a^2 - 2b^2 \neq 0$. Because

$$(a + b\sqrt{2})(a - b\sqrt{2}) = a^2 - 2b^2,$$

we see that $(a + b\sqrt{2})(a/c - \sqrt{2}\,b/c) = 1$, where $c = a^2 - 2b^2$. The required inverse for $a + b\sqrt{2}$ is $a/c - \sqrt{2}\,b/c$, which is certainly an element of F, since a/c and b/c are rational.

6. Let F be any field and let $F[x]$ be the ring of polynomials in x over F. Since $F[x]$ is an integral domain, it has a field of quotients according to Theorem 4.7.1, which consists of all quotients $f(x)/g(x)$, where $f(x)$ and $g(x)$ are in $F[x]$ and $g(x) \neq 0$. This field of quotients of $F[x]$ is denoted by $F(x)$ and is called the *field of rational functions in x over F.*

7. Z_p, the integers modulo the prime p, is a (finite) field.

8. In Example 2 in Section 4 of Chapter 4 we saw how to construct a field having nine elements.

These eight examples are specific ones. Using the theorems we have proved earlier, we have some general constructions of fields.

9. If D is any integral domain, then it has a field of quotients, by Theorem 4.7.1, which consists of all the fractions a/b, where a and b are in D and $b \neq 0$.

10. If R is a commutative ring with unit element 1 and M is a maximal ideal of R, then Theorem 4.4.2 tells us that R/M is a field.

This last example, for particular R's, will play an important role in what is to follow in this chapter.

We could go on, especially with special cases of Examples 9 and 10, to see more examples. But the 10 that we did see above show us a certain variety of fields, and we see that it is not too hard to run across fields.

In Examples 7 and 8 the fields are finite. If F is a finite field having q elements, viewing F merely as an abelian group under its addition, " $+$," we have, by Theorem 2.45, that $qx = 0$ for every $x \in F$. This is a behavior quite distinct from that which happens in the fields that we are used to, such as the rationals and reals.

We single out this kind of behavior in the

Definition.   A field F is said to have (or, to be of) *character-istic* $p \neq 0$ if for some positive integer p, $px = 0$ for all $x \in F$, and no positive integer smaller than p enjoys this property.

If a field F is not of characteristic $p \neq 0$ for any positive integer p, we call it a field of *characteristic* 0. So $Q, \mathbb{R}, \mathbb{C}$ are fields of characteristic 0, while Z_3 is of characteristic 3.

In the definition given above the use of the letter p for the character-istic is highly suggestive, for we have always used p to denote a prime. In fact, as we see in the next theorem, this usage of p is consistent.

Theorem 5.1.1.   The characteristic of a field is either 0 or a prime number.

Proof.   If the field F has characteristic 0, there is nothing more to say. Suppose then that $mx = 0$ for all $x \in F$, where m is a positive integer. Let p be the smallest positive integer such that $px = 0$ for all $x \in F$. We claim that p is a prime. For if $p = uv$, where $u > 1$ and $v > 1$ are integers, then in $F, (u1)(v1) = (uv)1 = 0$, where 1 is the unit element of F. But F, being a field, is an integral domain (Problem 1); therefore, $u1 = 0$ or $v1 = 0$. In either case we get that $0 = (u1)(x) = ux$ [or, similarly, $0 = (v1)x = vx$] for any x in F. But this contradicts our choice of p as the smallest integer with this property. Hence p is a prime. □

Note that we did not use the full force of the assumption that F was a field. We only needed that F was an integral domain (with 1). So if we define the characteristic of an integral domain to be 0 or the smallest positive integer p such that $px = 0$ for all $x \in F$, we obtain the same result. Thus the

Corollary.   If D is an integral domain, then its characteristic is either 0 or a prime number.

Problems

1. Show that a field is an integral domain.

2. Prove the Corollary even if D does not have a unit element.

3. Given a ring R, let $S = R[x]$ be the ring of polynomials in x over R, and let $T = S[y]$ be the ring of polynomials in y over S. Show that:
   (a) Any element $f(x, y)$ in T has the form $\Sigma\Sigma a_{ij}x^iy^j$, where the a_{ij} are in R.
   (b) In terms of the form of $f(x, y)$ in T given in Part (a), give the condition for the equality of two elements $f(x, y)$ and $g(x, y)$ in T.
   (c) In terms of the form for $f(x, y)$ in Part (a), give the formula for $f(x, y) + g(x, y)$, for $f(x, y)$, $g(x, y)$ in T.
   (d) Give the form for the product of $f(x, y)$ and $g(x, y)$ if $f(x, y)$ and $g(x, y)$ are in T. (T is called the *ring of polynomials in two variables over* R, and is denoted by $R[x, y]$.)

4. If D is an integral domain, show that $D[x, y]$ is an integral domain.

5. If F is a field and $D = F[x, y]$, the field of quotients of D is called the *field of rational functions in two variables over* F, and is usually denoted by $F(x, y)$. Give the form of the typical element of $F(x, y)$.

6. Prove that $F(x, y)$ is isomorphic to $F(y, x)$.

7. If F is a field of characteristic $p \neq 0$, show that $(a + b)^p = a^p + b^p$ for all $a, b \in F$. (**Hint:** Use the binomial theorem and the fact that p is a prime.)

8. If F is a field of characteristic $p \neq 0$, show that $(a + b)^m = a^m + b^m$, where $m = p^n$, for all a, b in F and any positive integer n.

9. Let F be a field of characteristic $p \neq 0$ and let $\varphi: F \rightarrow F$ be defined by $\varphi(a) = a^p$ for all $a \in F$.
   (a) Show that φ defines a monomorphism of F into itself.
   (b) Give an example of a field F where φ is *not* onto. (*Very hard*)

10. If F is a finite field of characteristic p, show that the mapping φ defined above is onto, hence is an automorphism of F.

2.   A BRIEF EXCURSION INTO VECTOR SPACES

To get into the things we should like to do in field theory, we need some technical equipment that we do not have as yet. This concerns the relation of two fields $K \supset F$ and what we would like to consider as some measure of the size of K compared to that of F. This size is what we shall call the *dimension* or *degree* of K over F.

However, in these considerations, much less is needed of K than that it be a field. We would be remiss if we proved these results only for the special context of two fields $K \supset F$ because the same ideas, proofs, and spirit hold in a far wider situation. We need the notion of a *vector space* over a field F. Aside from the fact that what we do in vector spaces will be important in our situation of fields, the ideas developed appear in all parts of mathematics. Students of algebra must see these things at some stage of their training. An appropriate place is right here.

> **Definition.**   A *vector space* V over a field F is an abelian group under " $+$ " such that for every $\alpha \in F$ and every $v \in V$ there is an element $\alpha v \in V$, and such that:
> 1. $\alpha(v_1 + v_2) = \alpha v_1 + \alpha v_2$, for $\alpha \in F$, $v_1, v_2 \in V$.
> 2. $(\alpha + \beta)v = \alpha v + \beta v$, for $\alpha, \beta \in F$, $v \in V$.
> 3. $\alpha(\beta v) = (\alpha\beta)v$, for $\alpha, \beta \in F$, $v \in V$.
> 4. $1v = v$ for all $v \in V$, where 1 is the unit element of F.

In discussing vector spaces—which we will do very briefly—we shall use lowercase Latin letters for elements of V and lowercase Greek letters for elements of F.

Our basic concern here will be with only one aspect of the theory of vector spaces: the notion of the dimension of V over F. We shall develop this notion as expeditiously as possible, not necessarily in the best or most elegant way. We would strongly advise the readers to see the other sides of what is done in vector spaces in other books on algebra or linear algebra (for instance, our books *A Primer on Linear Algebra* and *Matrix Theory and Linear Algebra*.)

Before getting down to some results, we look at some examples. We leave to the reader the details of verifying, in each case, that the example really is an example of a vector space.

EXAMPLES

1. Let F be any field and let $V = \{(\alpha_1, \alpha_2, \ldots, \alpha_n)| \text{ for all } i, \ \alpha_i \in F\}$ be the set of n-tuples over F, with equality and addition defined component-wise. For $v = (\alpha_1, \alpha_2, \ldots, \alpha_n)$ and $\beta \in F$, define βv by $\beta v = (\beta\alpha_1, \beta\alpha_2, \ldots, \beta\alpha_n)$. V is a vector space over F.

2. Let F be any field and let $V = F[x]$ be the ring of polynomials in x over F. Forgetting the product of any arbitrary elements of $F[x]$ but using only that of a polynomial by a constant, for example, we find that

$$\beta(\alpha_0 + \alpha_1 x + \cdots + \alpha_n x^n) = \beta\alpha_0 + \beta\alpha_1 x + \cdots + \beta\alpha_n x^n$$

V becomes a vector space over F.

3. Let V be as in Example 2 and let $W = \{f(x) \in V| \deg(f(x)) \leq n\}$. Then. W is a vector space over F, and $W \subset V$ is a subspace of V in the following sense.

> **Definition.** A *subspace* of a vector space V is a nonempty subset W of V such that $\alpha w \in W$ and $w_1 + w_2 \in W$ for all α in F and $w, w_1, w_2 \in W$.

Note that the definition of subspace W of V implies that W is a vector space whose operations are just those of V restricted to the elements of W.

4. Let V be the set of all real-valued differentiable functions on $[0, 1]$, the closed unit interval, with the usual addition and multiplication of a function by a real number. Then V is a vector space over $\mathbb{R}$.

5. Let W be all the real-valued continuous functions on $[0, 1]$, again with the usual addition and multiplication of a function by a real number. W, too, is a vector space over $\mathbb{R}$, and the V in Example 4 is a subspace of W.

6. Let F be any field, $F[x]$ the ring of polynomials in x over F. Let $f(x)$ be in $F[x]$ and $J = (f(x))$ the ideal of $F[x]$ generated by $f(x)$.

Let $V = F[x]/J$, where we define $\alpha(g(x) + J) = \alpha g(x) + J$. V is then a vector space over F.

7. Let $\mathbb{R}$ be the real field and let V be the set of all solutions to the differential equation $d^2y/dx^2 + y = 0$. V is a vector space over $\mathbb{R}$.

8. Let V be any vector space over a field F, $v_1, v_2, \ldots, v_n$ elements of V. Let $\langle v_1, v_2, \ldots, v_n \rangle = \{ \alpha_1 v_1 + \alpha_2 v_2 + \cdots + \alpha_n v_n | \alpha_1, \alpha_2, \ldots, \alpha_n \in F \}$. Then $\langle v_1, v_2, \ldots, v_n \rangle$ is a vector space over F and is a subspace of V. This subspace $\langle v_1, v_2, \ldots, v_n \rangle$ is called the *subspace of V generated or spanned by* $v_1, \ldots, v_n$ over F; its elements are called *linear combinations of* $v_1, \ldots, v_n$. We shall soon have a great deal to say about $\langle v_1, v_2, \ldots, v_n \rangle$.

9. Let V and W be vector spaces over the field F and let $V \oplus W = \{(v, w) | v \in V, w \in W)$, with equality and addition defined component-wise, and where $\alpha(v, w) = (\alpha v, \alpha w)$. Then $V \oplus W$ is easily seen to be a vector space over F; it is called the *direct sum* of V and W.

10. Let $K \supset F$ be two fields, with the addition, "$+$," that of K and where αv, for $\alpha \in F$ and $v \in K$ is the product, as elements of the field K. Then Conditions 1 and 2 defining a vector space are merely special cases of the distributive laws that hold in K, and Condition 3 is merely a consequence of the associativity of the product in K. Finally, Condition 4 is just the restatement of the fact that 1 is the unit element of K. So K is a vector space over F.

In one respect there is a sharp difference among these examples. We specify this difference by examining some of these examples in turn.

1. In Example 1, if

$$v_1 = (1, 0, \ldots, 0), \ v_2 = (0, 1, 0, \ldots, 0), \ldots, v_n = (0, 0, \ldots, 1),$$

then every element v in V has a *unique* representation in the form $v = \alpha_1 v_1 + \cdots + \alpha_n v_n$, where $\alpha_1, \ldots, \alpha_n$ are in F.
2. In Example 3, if $v_1 = 1, v_2 = x, \ldots, v_i = x^{i-1}, \ldots, v_{n+1} = x^n$, then every $v \in V$ has a *unique* representation as $v = \alpha_1 v_1 + \cdots + \alpha_n v_n$, with the α_i in F.
3. In Example 7, every solution of $d^2y/dx^2 + y = 0$ is of the unique form $y = \alpha \cos x + \beta \sin x$, with α and β real.

4. In Example 8, every $v \in \langle v_1, \ldots, v_n \rangle$ has a representation—*albeit not necessarily unique*—as $v = \alpha_1 v_1 + \cdots + \alpha_n v_n$ from the very definition of $\langle v_1, \ldots, v_n \rangle$. Uniqueness of this representation depends heavily on the elements $v_1, \ldots, v_n$.
5. In the special case of Example 10, where $K = \mathbb{C}$, the field of complex numbers, and $F = \mathbb{R}$ that of the real numbers, then every $v \in \mathbb{C}$ is of the *unique* form $v = \alpha + \beta i$, $\alpha, \beta \in \mathbb{R}$.
6. Consider $K = F(x) \supset F$, the field of rational functions in x over F. We claim—and leave to the reader—that we *cannot* find any finite set of elements in K which span K over F. This phenomenon was also true in some of the other examples we gave of vector spaces.

The whole focus of our attention here will be on this notion of a vector space having some finite subset that spans it over the base field.

Before starting this discussion, we must first dispose of a list of *formal* properties that hold in a vector space. You, dear reader, are by now so sophisticated in dealing with these formal, abstract things that we leave the proof of the next lemma to you.

Lemma 5.2.1. If V is a vector space over the field F, then, for every $\alpha \in F$ and every $v \in V$:
    (a) $\alpha 0 = 0$, where 0 is the zero-element of V.
    (b) $0v = 0$, where 0 is the zero in F.
    (c) $\alpha v = 0$ implies that $\alpha = 0$ or $v = 0$.
    (d) $(-\alpha)v = -(\alpha v)$.

In view of this lemma we shall not run into any confusion if we use the symbol 0 both for the zero of F and that of V.

We forget vector spaces for a moment and look at solutions of certain systems of linear equations in fields. Take, for example, the two linear homogeneous equations with real coefficients, $x_1 + x_2 + x_3 = 0$ and $3x_1 - x_2 + x_3 = 0$. We easily see that for any x_1, x_3 such that $4x_1 + 2x_3 = 0$ and $x_2 = -(x_1 + x_3)$, we get a solution to this system. In fact, there exists an infinity of solutions to this system other than the trivial one $x_1 = 0$, $x_2 = 0$, $x_3 = 0$. If we look at this example and ask ourselves: Why is there an infinity of solutions to this system of linear equations? We quickly come to the conclusion that because there are more variables than equations we have room to maneuver to produce solutions. This is exactly the situation that holds more generally, as we see below.

> **Definition.**   Let F be a field; then the n-tuple $(\beta_1, \ldots, \beta_n)$, where the β_i are in F, and *not all of them are* 0, is said to be a *nontrivial solution in F* to the system of homogeneous linear equations

$$\alpha_{11}x_1 + \alpha_{12}x_2 + \cdots + \alpha_{1n}x_n = 0$$

$$\alpha_{21}x_1 + \alpha_{22}x_2 + \cdots + \alpha_{2n}x_n = 0$$

$$\cdot \qquad \cdot \qquad \cdots \qquad \cdot \qquad = 0$$

(*) $\qquad \cdot \qquad \cdot \qquad \cdots \qquad \cdot \qquad = 0$

$$\alpha_{i1}x_1 + \alpha_{i2}x_2 + \cdots + \alpha_{in}x_n = 0$$

$$\cdot \qquad \cdot \qquad \cdots \qquad \cdot \qquad = 0$$

$$\alpha_{r1}x_1 + \alpha_{r2}x_2 + \cdots + \alpha_{rn}x_n = 0$$

> where the α_{ij} are all in F, if substituting $x_1 = \beta_1, \ldots, x_n = \beta_n$ satisfies all the equations of (*).

For such a system (*) we have the following

> **Theorem 5.2.2.**   If $n > r$, that is, if the number of variables (unknowns) exceeds the number of equations in (*), then (*) has a nontrivial solution in F.

Proof.   The method is that, which some of us learned in high school, of solving simultaneous equations by eliminating one of the unknowns and at the same time cutting the number of equations down by one.

We proceed by induction on r, the number of equations. If $r = 1$, the system (*) reduces to $\alpha_{11}x_1 + \cdots + \alpha_{1n}x_n = 0$, and $n > 1$. If all the $\alpha_{1i} = 0$, then $x_1 = x_2 = \cdots = x_n = 1$ is a nontrivial solution (*). So, on renumbering, we may assume that $\alpha_{11} \neq 0$; we then have the solution to (*), which is nontrivial: $x_2 = \cdots = x_n = 1$ and $x_1 = -(1/\alpha_{11})(\alpha_{12} + \cdots + \alpha_{1n})$.

Suppose that the result is correct for $r = k$ for some k and suppose that (*) is a system of $k + 1$ linear homogeneous equations in $n > k + 1$ variables. As above, we may assume that some $\alpha_{ij} \neq 0$, and, without loss of generality, that $\alpha_{11} \neq 0$.

We construct a related system, (**), of k linear homogeneous equations in $n - 1$ variables; since $n > k + 1$, we have that $n - 1 > k$, so we can apply induction to this new system (**). How do we get this new system? We want to eliminate x_1 among the equations. We do so by subtracting α_{i1}/α_{11} times the first equation from the ith one for each of $i = 2, 3, \ldots, k + 1$. In doing so, we end up with the new system of k linear homogeneous equations in $n - 1$ variables:

$$\beta_{22}x_2 \quad + \cdots + \beta_{2n}x_n \quad = 0$$

$$\beta_{32}x_2 \quad + \cdots + \beta_{3n}x_n \quad = 0$$

(**)
$$\qquad . \qquad \cdots \quad .$$

$$\qquad . \qquad \cdots \quad .$$

$$\beta_{k+1,2}x_2 + \cdots + \beta_{k+1,n}x_n = 0,$$

where $\beta_{ij} = \alpha_{ij} - \alpha_{i1}/\alpha_{11}$ for $i = 2, 3, \ldots, k + 1$ and $j = 2, 3, \ldots, n$.

Since (**) is a system of k linear homogeneous equations in $n - 1$ variables and $n - 1 > k$, by our induction (**) has a nontrivial solution $(\gamma_2, \ldots, \gamma_n)$ in F. Let $\gamma_1 = -(\alpha_{12}\gamma_2 + \cdots + \alpha_{1n}\gamma_n)/\alpha_{11}$; we leave it to the reader to verify that the $(\gamma_1, \gamma_2, \ldots, \gamma_n)$ so obtained is a required nontrivial solution to (*). This completes the induction and so proves the theorem. $\square$

With this result established, we are free to use it in our study of vector spaces. We now return to these spaces. We repeat, for emphasis, something we defined earlier in Example 8.

Definition. Let V be a vector space over F and let $v_1, v_2, \ldots, v_n$ be in V. The element $v \in V$ is said to be a *linear combination* of $v_1, v_2, \ldots, v_n$ if $v = \alpha_1 v_1 + \cdots + \alpha_n v_n$ for some $\alpha_1, \cdots \alpha_n$ in F.

As we indicated in Example 8, the set $\langle v_1, v_2, \ldots, v_n \rangle$ of all linear combinations of $v_1, v_2, \ldots, v_n$ is a vector space over F, and being contained in V, is a subspace of V. Why is it a vector space? If $\alpha_1 v_1 + \cdots + \alpha_n v_n$ and $\beta_1 v_1 + \cdots + \beta_n v_n$ are two linear combinations of $v_1, \ldots, v_n$, then

$$(\alpha_1 v_1 + \cdots + \alpha_n v_n) + (\beta_1 v_1 + \cdots + \beta_n v_n)$$
$$= (\alpha_1 + \beta_1) v_1 + \cdots + (\alpha_n + \beta_n) v_n$$

by the axioms defining a vectors space, and so is in $\langle v_1, \ldots, v_n \rangle$. If $\gamma \in F$ and $\alpha_1 v_1 + \cdots + \alpha_n v_n \in \langle v_1, \ldots, v_n \rangle$, then

$$\gamma(\alpha_1 v_1 + \cdots + \alpha_n v_n) = \gamma \alpha_1 v_1 + \cdots + \gamma \alpha_n v_n,$$

so is also in $\langle v_1, \ldots, v_n \rangle$. Thus $\langle v_1, \ldots, v_n \rangle$ is a vector space. As we called it earlier, it is the subspace of V *spanned* over F by $v_1, \ldots, v_n$.

This leads us to the ultra-important

Definition. The vector space V over F is *finite dimensional* over F if $V = \langle v_1, \ldots, v_n \rangle$ for some $v_1, \ldots, v_n$ in V, that is, if V is spanned over F by a finite set of elements.

Otherwise, we say that V is *infinite dimensional* over F if it is not finite dimensional over F. Note that although we have defined what is meant by a finite-dimensional vector space, we still have not defined what is meant by its dimension. That will come in due course.

Suppose that V is a vector space over F and $v_1, \ldots, v_n$ in V are such that every element v in $\langle v_1, \ldots, v_n \rangle$ has a *unique* representation in the form $v = \alpha_1 v_1 + \cdots + \alpha_n v_n$, where $\alpha_1, \ldots, \alpha_n \in F$. Since

$$0 \in \langle v_1, \ldots, v_n \rangle \quad \text{and} \quad 0 = 0 v_1 + \cdots + 0 v_n,$$

by the uniqueness we have assumed we obtain that if $\alpha_1 v_1 + \cdots + \alpha_n v_n = 0$, then $\alpha_1 = \alpha_2 = \cdots = \alpha_n = 0$. This prompts a second ultra-important definition, that which is singled out in the

> **Definition.** Let V be a vector space over F; then the elements $v_1, \ldots, v_n$ in V are said to be *linearly independent over* F if $\alpha_1 v_1 + \cdots + \alpha_n v_n = 0$, where $\alpha_1, \ldots, \alpha_n$ are in F, implies that $\alpha_1 = \alpha_2 = \cdots = \alpha_n = 0$.

If the elements $v_1, \ldots, v_n$ in V are *not* linearly independent over F, then we say that they are *linearly dependent over* F. For example, if $\mathbb{R}$ is the field of real numbers and V is the set of 3-tuples over $\mathbb{R}$ as defined in Example 1, then $(0, 0, 1)$, $(0, 1, 0)$, and $(1, 0, 0)$ are linearly independent over $\mathbb{R}$ (Prove!) while $(1, -2, 7)$, $(0, 1, 0)$, and $(1, -3, 7)$ are linearly dependent over $\mathbb{R}$, since $1(1, -2, 7) + (-1)(0, 1, 0) + (-1)(1, -3, 7) = (0, 0, 0)$ is a nontrivial linear combination of these elements over $\mathbb{R}$, which is the 0-vector.

Note that linear independence depends on the field F. If $\mathbb{C} \supset \mathbb{R}$ are the complex and real fields, respectively, then $\mathbb{C}$ is a vector space over $\mathbb{R}$, but it is also a vector space over $\mathbb{C}$ itself. The elements $1, i$ in $\mathbb{C}$ are linearly independent over $\mathbb{R}$ but are not so over $\mathbb{C}$, since $i1 + (-1)i = 0$ is a nontrivial linear combination of $1, i$ over $\mathbb{C}$.

We prove

> **Lemma 5.2.3.** If V is a vector space over F and $v_1, \ldots, v_n$ in V are linearly independent over F, then every element $v \in \langle v_1, \ldots, v_n \rangle$ has a unique representation as
>
> $$v = \alpha_1 v_1 + \cdots + \alpha_n v_n$$
>
> with $\alpha_1, \ldots, \alpha_n$ in F.

Proof. Suppose that $v \in \langle v_1, \ldots, v_n \rangle$ has the two representations as $v = \alpha_1 v_1 + \cdots + \alpha_n v_n = \beta_1 v_1 + \cdots + \beta_n v_n$ with the α's and β's in F. This gives us that $(\alpha_1 - \beta_1)v_1 + \cdots + (\alpha_n - \beta_n)v_n = 0$; since $v_1, \ldots, v_n$ are linearly independent over F, we conclude that $\alpha_1 - \beta_1 = 0, \ldots, \alpha_n - \beta_n = 0$, yielding for us the uniqueness of the representation. $\square$

How finite is a finite-dimensional vector space? To measure this, call a subset $v_1, \ldots, v_n$ of V a *minimal generating set* for V over F if $V = \langle v_1, \ldots, v_n \rangle$ and no set of fewer than n elements spans V over F. We now come to the third vitally important.

> **Definition.** If V is a finite-dimensional vector space over F, then the *dimension of V over F*, written $\dim_F(V)$, is n, the number of elements in a minimal generating set for V over F.

In the examples given, $\dim_{\mathbf{R}}(\mathbf{C}) = 2$, since $1, i$ is a minimal generating set for $\mathbf{C}$ over $\mathbf{R}$. However, $\dim_{\mathbf{C}}(\mathbf{C}) = 1$. In Example 1, $\dim_F(V) = n$ and in Example 3, $\dim_F(V) = n + 1$. In Example 7 the dimension of V over F is 2. Finally, if $\langle v_1, \ldots, v_n \rangle \subset V$, then $\dim_F \langle v_1, \ldots, v_n \rangle$ is *at most n*.

We now prove

> **Lemma 5.2.4.** If V is finite dimensional over F of dimension n and if the elements $v_1, \ldots, v_n$ of V generate V over F, then $v_1, \ldots, v_n$ are linearly independent over F.

Proof. Suppose that $v_1, \ldots, v_n$ are linearly dependent over F; thus there is a linear combination $\alpha_1 v_1 + \cdots + \alpha_n v_n = 0$, where not all the α_i are 0. We may suppose, without loss of generality, that $\alpha_1 \neq 0$; then $v_1 = (-1/\alpha_1)(\alpha_2 v_2 + \cdots + \alpha_n v_n)$. Given $v \in V$, because $v_1, \ldots, v_n$ is a generating set for V over F,

$$v = \beta_1 v_1 + \cdots + \beta_n v_n = \left(-\frac{\beta_1}{\alpha_1}\right)(\alpha_2 v_2 + \cdots + \alpha_n v_n) + \beta_2 v_2$$

$$+ \cdots + \beta_n v_n;$$

thus $v_2, \ldots, v_n$ span V over F, contradicting that the subset $v_1, v_2, \ldots, v_n$ is a minimal generating set of V over F. $\square$

We now come to yet another important

Definition. Let V be a finite-dimensional vector space over F; then $v_1, \ldots, v_n$ is a *basis* of V over F if the elements $v_1, \ldots, v_n$ span V over F and are linearly independent over F.

By Lemma 5.2.4 any minimal generating set of V over F is a basis of V over F. Thus finite-dimensional vector spaces have bases. We proceed to

Theorem 5.2.5. Suppose that V is finite dimensional over F; then any two bases of V over F have the same number of elements, and this number is exactly $\dim_F(V)$.

Proof. Let $v_1, \ldots, v_n$ and $w_1, \ldots, w_m$ be two bases of V over F. We want to show that $m = n$. Suppose that $m > n$. Because $v_1, \ldots, v_n$ is a basis of V over F, we know that every element in V is a linear combination of the v_i over F. In particular, $w_1, \ldots, w_m$ are each a linear combination of $v_1, \ldots, v_n$ over F. Thus we have

$$w_1 = \alpha_{11}v_1 + \alpha_{12}v_2 + \cdots + \alpha_{1n}v_n$$

$$w_2 = \alpha_{21}v_1 + \alpha_{22}v_2 + \cdots + \alpha_{2n}v_n$$

$$\vdots \qquad\qquad \vdots$$

$$w_m = \alpha_{m1}v_1 + \alpha_{m2}v_2 + \cdots + \alpha_{mn}v_n$$

where the α_{ij} are in F.
    Consider

$$\beta_1 w_1 + \cdots + \beta_m w_m = (\alpha_{11}\beta_1 + \alpha_{21}\beta_2 + \cdots + \alpha_{m1}\beta_m)v_1$$

$$+ \cdots + (\alpha_{1n}\beta_1 + \alpha_{2n}\beta_2 + \cdots + \alpha_{mn})v_n.$$

The system of linear homogeneous equations

$$\alpha_{1i}\beta_1 + \alpha_{2i}\beta_2 + \cdots + \alpha_{mi}\beta_m = 0, \ i = 1, 2, \ldots, n,$$

has a nontrivial solution in F by Theorem 5.2.2, since the number of variables, m, exceeds the number of equations, n. If $\beta_1, \ldots, \beta_m$ is such a solution in F, then, by the above, $\beta_1 w_1 + \cdots + \beta_m w_m = 0$, yet not all the β_i are 0. This contradicts the linear independence of $w_1, \ldots, w_m$ over F. Therefore, $m \leq n$. Similarly, $n \leq m$; hence $m = n$. The theorem is now proved, since a minimal generating set of V over F is a basis of V over F and the number of elements in this minimal generating set is $\dim_F(V)$, by definition. Therefore, by the above, $n = \dim_F(V)$, completing the proof. $\square$

A further result, which we shall use in field theory, of a similar nature to the things we have been doing is

> **Theorem 5.2.6.** Let V be a vector space over F such that $\dim_F(V) = n$. If $m > n$, then any m elements of V are linearly dependent over F.

Proof. Let $w_1, \ldots, w_m \in V$ and let $v_1, \ldots, v_n$ be a basis of V over F; here $n = \dim_F(V)$ by Theorem 5.2.5. Therefore,

$$w_1 = \alpha_{11} v_1 + \cdots + \alpha_{1n} v_n, \ldots, w_m = \alpha_{m1} v_1 + \cdots + \alpha_{mn} v_n.$$

The proof given in Theorem 5.2.5, that if $m > n$ we can find $\beta_1, \ldots \beta_m$ in F, and not all 0, such that $\beta_1 w_1 + \cdots + \beta_m w_m = 0$, goes over word for word. But this establishes that $w_1, \ldots, w_m$ are linearly dependent over F. $\square$

We close this section with a final theorem of the same flavor as the preceding ones.

> **Theorem 5.2.7.** Let V be a vector space over F with $\dim_F(V) = n$. Then any n linearly independent elements of V form a basis of V over F.

Proof. We want to show that if $v_1, \ldots, v_n \in V$ are linearly independent over F, then they span V over F. Let $v \in V$; then $v, v_1, \ldots, v_n$ are $n + 1$ elements, hence, by Theorem 5.2.6, they are linearly dependent over F.

Thus there exist elements $\alpha, \alpha_1, \ldots, \alpha_n$ in F, not all 0, such that $\alpha v + \alpha_1 v_1 + \cdots + \alpha_n v_n = 0$. The element α *cannot* be 0, otherwise $\alpha_1 v_1 + \cdots + \alpha_n v_n = 0$, and not all the α_i are 0. This would contradict the linear independence of the elements $v_1, \ldots, v_n$ over F. Thus $\alpha \neq 0$, and so $v = (-1/\alpha)(\alpha_1 v_1 + \cdots + \alpha_n v_n) = \beta_1 v_1 + \cdots + \beta_n v_n$, where $\beta_i = -\alpha_i/\alpha_1$. Therefore, $v_1, \ldots, v_n$ span V over F, and thus must form a basis of V over F. $\square$

Problems

Easier Problems

1. Determine if the following elements in V, the vector space of 3-tuples over $\mathbb{R}$, are linearly independent over $\mathbb{R}$.
   (a) $(1, 2, 3), (4, 5, 6), (7, 8, 9)$.
   (b) $(1, 0, 1), (0, 1, 2), (0, 0, 1)$.
   (c) $(1, 2, 3), (0, 4, 5), (\frac{1}{2}, 3, \frac{21}{4})$.

2. Find a nontrivial solution in Z_5 of the system of linear homogeneous equations:

$$x_1 + x_2 + x_3 = 0$$

$$x_1 + 2x_2 + 3x_3 = 0$$

$$3x_1 + 4x_2 + 2x_3 = 0.$$

3. If V is a vector space of dimension n over Z_p, p a prime, show that V has p^n elements.

4. Prove all of Lemma 5.2.1.

5. Let F be a field and $V = F[x]$, the polynomial ring in x over F. Considering V as a vector space over F, prove that V is not finite-dimensional over F.

6. If V is a finite-dimensional vector space over F and if W is a subspace of V, prove that:
   (a) W is finite dimensional over F and $\dim_F(W) \leq \dim_F(V)$.
   (b) if $\dim_F(W) = \dim_F(V)$, then $V = W$.

*7. Define what you feel should be a vector space homomorphism ψ of V into W, where V and W are vector spaces over F. What can you say about the kernel, K, of ψ where $K = \{v \in V | \psi(v) = 0\}$? What should a vector space isomorphism be?

8. If V is a vector space over F and W is a subspace of V, define the requisite operations in V/W so that V/W becomes a vector space over F.

9. Show that if $\dim_F(V) = n$ and W is a subspace of V with $\dim_F(W) = m$, then $\dim_F(V/W) = n - m$.

10. If $\psi: V \to V'$ is a homomorphism of V *onto* V' with kernel K, show that $V' \simeq V/K$ (as vector spaces over F). (See Problem 7).

11. If V is a finite-dimensional vector space over F and $v_1, \ldots, v_m$ in V are linearly independent over F, show we can find $w_1, \ldots, w_r$ in V, where $m + r = \dim_F(V)$, such that $v_1, \ldots, v_m, w_1, \ldots, w_r$ form a basis of V over F.

12. If V is a vector space over F of dimension n, prove that V is isomorphic to the vector space of n-tuples over F (Example 1). (See Problem 7).

Middle-Level Problems

13. Let $K \supset F$ be two fields; suppose that K, as a vector space over F, has finite dimension n. Show that if $a \in K$, then there exist $\alpha_0, \alpha_1, \ldots, \alpha_n$ in F, not all 0, such that

$$\alpha_0 + \alpha_1 a + \alpha_2 a^2 + \cdots + \alpha_n a^n = 0.$$

14. Let F be a field, $F[x]$ the polynomial ring in x over F, and $f(x) \neq 0$ in $F[x]$. Consider $V = F[x]/J$ as a vector space over F, where J is the ideal of $F[x]$ generated by $f(x)$. Prove that

$$\dim_F(V) = \deg f(x).$$

15. If V and W are two finite-dimensional vector spaces over F, prove that $V \oplus W$ is finite dimensional over F and that $\dim_F(V \oplus W) = \dim_F(V) + \dim_F(W)$.

16. Let V be a vector space over F and suppose that U and W are subspaces of V. Define $U + W = \{u + w | u \in U, w \in W\}$. Prove that:
    (a) $U + W$ is a subspace of V.
    (b) $U + W$ is finite dimensional over F if both U and W are.
    (c) $U \cap W$ is a subspace of V.
    (d) $U + W$ is a homomorphic image of $U \oplus W$.
    (e) If U and W are finite dimensional over F, then

$$\dim_F (U + W) = \dim_F (U) + \dim_F (W) - \dim_F (U \cap W).$$

Harder Problems

17. Let $K \supset F$ be two fields such that $\dim_F (K) = m$. Suppose that V is a vector space over K. Prove that:
    (a) V is a vector space over F.
    (b) If V is finite dimensional over K, then it is finite dimensional over F.
    (c) If $\dim_K (V) = n$, then $\dim_F (V) = mn$ [i.e., $\dim_F (V) = \dim_K (V) \dim_F (K)$].

18. Let $K \supset F$ be fields and suppose that V is a vector space over K such that $\dim_F (V)$ is finite. If $\dim_F (K)$ is finite, show that $\dim_K (V)$ is finite and determine its value in terms of $\dim_F (V)$ and $\dim_F (K)$.

19. Let D be an integral domain with 1, which happens to be a finite-dimensional vector space over a field F. Prove that D is a field. (**Note:** Since $F1$, which we can identify with F, is in D, the ring structure of D and the vector space structure of D over F are in harmony with each other.)

20. Let V be a vector space over an *infinite* field F. Show that V *cannot* be the set-theoretic union of a *finite* number of *proper* subspaces of V. (*Very hard*)

3. FIELD EXTENSIONS

Our attention now turns to a relationship between two fields K and F, where $K \supset F$. We call K an *extension* (or *extension field*) of F, and call

F a *subfield* of *K*. The operations in *F* are just those of *K* restricted to the elements of *F*. *In all that follows in this section it will be understood that* $K \subset F$.

We say that *K* is a *finite extension* of *F* if, viewed as a vector space over *F*, $\dim_F(K)$ is finite. We shall write $\dim_F(K)$ as $[K : F]$ and call it the *degree of K over F*.

We begin our discussion with what is usually the first result one proves in talking about finite extensions.

> **Theorem 5.3.1.** Let $L \supset K \supset F$ be three fields such that both $[L : K]$ and $[K : F]$ are finite. Then *L* is a finite extension of *F* and $[L : F] = [L : K][K : F]$.

Proof. We shall prove that *L* is a finite extension of *F* by explicitly exhibiting a finite basis of *L* over *F*. In doing so, we shall obtain the stronger result asserted in the theorem, namely that $[L : F] = [L : K][K : F]$.

Suppose that $[L : K] = m$ and $[K : F] = n$; then *L* has a basis $v_1, v_2, \ldots, v_m$ over *K*, and *K* has a basis $w_1, w_2, \ldots, w_n$ over *F*. We shall prove that the *mn* elements $v_i w_j$, where $i = 1, 2, \ldots, m$ and $j = 1, 2, \ldots, n$, constitute a basis of *L* over *F*.

We begin by showing that, at least, these elements span *L* over *F*; this will, of course, show that *L* is a finite extension of *F*. Let $a \in L$; since the elements $v_1, \ldots, v_m$ form a basis of *L* over *K*, we have $a = k_1 v_1 + \cdots + k_m v_m$, where $k_1, k_2, \ldots, k_m$ are in *K*. Since $w_1, \ldots, w_n$ is a basis of *K* over *F*, we can express each k_i as

$$k_i = f_{i1} w_1 + f_{i2} w_2 + \cdots + f_{in} w_n,$$

where the f_{ij} are in *F*. Substituting these expressions for the k_i in the foregoing expression of *a*, we obtain

$$a = (f_{11} w_1 + f_{12} w_2 + \cdots + f_{1n} w_n) v_1$$

$$+ \cdots + (f_{m1} v_1 + f_{m2} v_2 + \cdots + f_{mn} v_n) v_m.$$

Therefore, on unscrambling this sum explicitly, we obtain that

$$a = f_{11}w_1v_1 + f_{12}w_2v_1 + \cdots + f_{ij}w_jv_i + \cdots + f_{mn}w_nv_m.$$

Thus the mn elements v_iw_j in L span L over F; therefore, $[L:F]$ is finite and, in fact, $[L:F] \le mn$.

To show that $[L:F] = mn$, we need only show that the mn elements v_iw_j above are linearly independent over F, for then—together with the fact that they span L over F—we would have that they form a basis of L over F. By Theorem 5.2.5 we would have the desired result $[L:F] = mn = [L:K][K:F]$.

Suppose then that for some b_{ij} in F we have the relation

$$0 = b_{11}v_1w_1 + b_{12}v_1w_1 + \cdots + b_{1n}v_1w_n + b_{21}v_2w_1$$

$$+ \cdots + b_{2n}v_2w_n + \cdots + b_{m1}v_mw_1 + \cdots + b_{mn}v_mw_n.$$

Reassembling this sum, we obtain that $c_1v_1 + c_2v_2 + \cdots + c_mv_m = 0$, where $c_1 = b_{11}w_1 + \cdots + b_{1n}w_n, \ldots, c_m = b_{m1}w_1 + \cdots + b_{mn}w_n$. Since the c_i are elements of K and the elements $v_1, \ldots, v_n$ in L are linearly independent over K, we obtain that $c_1 = c_2 = \cdots = c_m = 0$.

Recalling that $c_i = b_{i1}w_1 + \cdots + b_{in}w_n$, where the b_{ij} are in F and where $w_1, \ldots, w_n$ in K are linearly independent over F, we deduce from the fact that $c_1 = c_2 = \cdots = c_m = 0$ that every $b_{ij} = 0$. Thus only the trivial linear combination, with each coefficient 0, of the elements v_iw_j over F can be 0. Hence the v_iw_j are linearly independent over F. We saw above that this was enough to prove the theorem. $\square$

The reader should compare Theorem 5.3.1 with the slightly more general result in Problem 17 of Section 2. The reader should now be able to solve Problem 17.

As a consequence of the theorem we have the

Corollary. If $L \supset K \supset F$ are three fields such that $[L:F]$ is finite, then $[K:F]$ is finite and divides $[L:F]$.

Proof. Since $L \supset K$, K cannot have more linearly independent elements over F than does L. Because, by Theorem 5.2.6, $[L:F]$ is the size of the largest set of linearly independent elements in L over F, we therefore get that $[K : F] \leq [L : F]$, so must be finite. Since L is finite dimensional over F and since K contains F, L must be finite dimensional over K. Thus all the conditions of Theorem 5.3.1 are fulfilled, whence $[L : F] = [L : K][K : F]$. Consequently, $[K : F]$ divides $[L : F]$, as is asserted in the Corollary. $\square$

If K is a finite extension of F, we can say quite a bit about the behavior of the elements of K vis-à-vis F.

Theorem 5.3.2. Suppose that K is a finite extension of F of degree n. Then, given any element u in K there exist elements $\alpha_0, \alpha_1, \ldots, \alpha_n$ in F, not all zero, such that

$$\alpha_0 + \alpha_1 u + \cdots + \alpha_n u^n = 0.$$

Proof. Since $[K : F] = \dim_F(K) = n$ and the elements $1, u, u^2, \ldots, u^n$ are $n + 1$ in number, by Theorem 5.2.6 they must be linearly dependent over F. Thus we can find $\alpha_0, \alpha_1, \ldots, \alpha_n$ in F, with not all of them 0, such that $\alpha_0 + \alpha_1 u + \alpha_2 u^2 + \cdots + \alpha_n u^n = 0$, proving the theorem. $\square$

The conclusion of the theorem suggests that we single out elements in an extension field that satisfy a nontrivial polynomial.

Definition. If $K \supset F$ are fields, then $a \in K$ is said to be *algebraic over* F if there exists a polynomial $p(x) \neq 0$ in $F[x]$ such that $p(a) = 0$.

By $p(a)$ we shall mean the element $\alpha_0 a^n + \alpha_1 a^{n-1} + \cdots + \alpha_n$ in K, where $p(x) = \alpha_0 x^n + \alpha_1 x^{n-1} + \cdots + \alpha_n$.

If K is an extension of F such that every element of K is algebraic over F, we call K an *algebraic extension* of F. In these terms Theorem 5.3.2 can be restated as: *If K is a finite extension of F, then K is an algebraic extension of F.*

The converse of this is not true; an algebraic extension of F *need not* be of finite degree over F. Can you come up with an example of this situation?

An element of K that is not algebraic over F is said to be *transcendental* over F.

Let's see some examples of algebraic elements in a concrete context. Consider $\mathbf{C} \supset \mathbf{Q}$, the complex field as an extension of the rational one. The complex number $a = 1 + i$ is algebraic over Q, since it satisfies $a^2 - 2a + 2 = 0$ over $\mathbf{Q}$. Similarly, the real number $b = \sqrt{1 + \sqrt[3]{1 + \sqrt{2}}}$ is algebraic over Q, since $b^2 = 1 + \sqrt[3]{1 + \sqrt{2}}$, so $(b^2 - 1)^3 = 1 + \sqrt{2}$, and therefore $((b^2 - 1)^3 - 1)^2 = 2$. Expanding this out, we get a non-trivial polynomial expression in b with rational coefficients, which is 0. Thus b is algebraic over $\mathbf{Q}$.

It is possible to construct real numbers that are transcendental over $\mathbf{Q}$ fairly easily (see Section 6 of Chapter 6). However, it takes some real effort to establish the transcendence of certain familiar numbers. The two familiar numbers e and π can be shown to be transcendental over $\mathbf{Q}$. That e is such was proved by Hermite in 1873; the proof that π is transcendental over $\mathbf{Q}$ is much harder and was first carried out by Lindemann in 1882. We shall not go into the proof here that any particular number is transcendental over $\mathbf{Q}$. However, in Section 7 of Chapter 6 we shall at least show that π is irrational. This makes it a possible candidate for a transcendental number of $\mathbf{Q}$, for clearly any rational number b is algebraic over $\mathbf{Q}$ because it satisfies the polynomial $p(x) = x - b$, which has rational coefficients.

Definition. A complex number is said to be an *algebraic number* if it is algebraic over $\mathbf{Q}$.

As we shall soon see, the algebraic numbers form a field, which is a subfield of $\mathbf{C}$.

We return to the general development of the theory of fields. We have seen in Theorem 5.3.2 that if K is a finite extension of F, then every element of K is algebraic over F. We turn this matter around by asking: If K is an extension of F and $a \in K$ is algebraic over F, can we somehow produce a finite extension of F using a?. The answer is yes. This will come as a consequence of the next theorem—which we prove in a context a little more general than what we really need.

Theorem 5.3.3.   Let D be an integral domain with 1 which is a finite-dimensional vector space over a field F. Then D is a field.

Proof.   To prove the theorem, we must produce for $a \neq 0$ in D an inverse, a^{-1}, in D such that $aa^{-1} = 1$.

As in the proof of Theorem 5.3.2, if $\dim_F (D) = n$, then $1, a, a^2, \ldots, a^n$ in D are linearly dependent over F. Thus for some appropriate $\alpha_0, \alpha_1, \ldots, \alpha_n$ in F, not all of which are 0,

$$\alpha_0 a^n + \alpha_1 a^{n-1} + \cdots + \alpha_n = 0.$$

Let $p(x) = \beta_0 x^r + \beta_1 x^{r-1} + \cdots + \beta_r \neq 0$ be a polynomial in $F[x]$ of lowest degree such that $p(a) = 0$. We assert that $\beta_r \neq 0$. For if $\beta_r = 0$, then

$$0 = \beta_0 a^r + \beta_1 a^{r-1} + \cdots + \beta_{r-1} a$$

$$= \left(\beta_0 a^{r-1} + \beta_1 a^{r-2} + \cdots + \beta_{r-1} \right) a;$$

since D is an integral domain and $a \neq 0$, we conclude that $\beta_0 a^{r-1} + \beta_1 a^{r-2} + \cdots + \beta_{r-1} = 0$, hence $q(a) = 0$, where $q(x) = \beta_0 x^{r-1} + \beta_1 x^{r-2} + \cdots + \beta_{r-1}$ in $F[x]$ is of lower degree than $p(x)$, a contradiction. Thus $\beta_r \neq 0$, hence β_r^{-1} is in F and

$$\frac{a \left(\beta_0 a^{r-1} + \cdots + \beta_{r-1} \right)}{\beta_r} = -1,$$

giving us that $-(\beta_0 a^{r-1} + \cdots + \beta_{r-1})/\beta_r$, which is in D, is the a^{-1} in D that we required. This proves the theorem. $\square$

Having Theorem 5.3.3 in hand, we want to make use of it. So, how do we produce subrings of a field K that contains F and are finite dimensional over F? Such subrings, as subrings of a field, are automatically integral domains, and would satisfy the hypothesis of Theorem

5.3.3. The means to this end will be the elements in K that are algebraic over F.

But first a definition.

> **Definition.** The element a in the extension K of F is said to be *algebraic of degree n* if there is a polynomial $p(x)$ in $F[x]$ of degree n such that $p(a) = 0$, and no nonzero polynomial of lower degree in $F[x]$ has this property.

We may assume that the polynomial $p(x)$ in this definition is monic, for we could divide this polynomial by its highest coefficient to obtain a monic polynomial $q(x)$ in $F[x]$, of the same degree as $p(x)$, and such that $q(a) = 0$. We henceforth assume that this polynomial $p(x)$ is monic; we call it the *minimal polynomial for a over F*.

> **Lemma 5.3.4.** Let $a \in K$ be algebraic over F with minimal polynomial $p(x)$ in $F[x]$. Then $p(x)$ is irreducible in $F[x]$.

Proof. Suppose that $p(x)$ is not irreducible in $F[x]$; then $p(x) = f(x)g(x)$ where $f(x)$ and $g(x)$ are in $F[x]$ and each has *positive* degree. Since $0 = p(a) = f(a)g(a)$, and since $f(a)$ and $g(a)$ are in the field K, we conclude that $f(a) = 0$ or $g(a) = 0$, both of which are impossible, since both $f(x)$ and $g(x)$ are of lower degree than $f(x)$. Therefore, $p(x)$ is irreducible in $F[x]$. $\square$

Let $a \in K$ be algebraic of degree n over F and let $p(x) \in F[x]$ be its minimal polynomial over F. Given $f(x) \in F[x]$, then $f(x) = q(x)p(x) + r(x)$, where $q(x)$ and $r(x)$ are in $F[x]$ and $r(x) = 0$ or $\deg r(x) < \deg p(x)$ follows from the division algorithm. Therefore, $f(a) = q(a)p(a) + r(a) = r(a)$, since $p(a) = 0$. In short, any polynomial expression in a over F can be expressed as a polynomial expression in a of degree *at most* $n - 1$.

Let $F[a] = \{ f(a) | f(x) \in F[x] \}$. We claim that $F[a]$ is a subfield of K that contains both F and a, and that $[F[a] : F] = n$. By the remark made above, $F[a]$ is spanned over F by $1, a, a^2, \ldots, a^{n-1}$, so is finite dimensional over F. Moreover, as is easily verified, $F[a]$ is a subring of

K; as a subring of K, $F[a]$ is an integral domain. Thus, by Theorem 5.3.3, $F[a]$ is a field. Since it is spanned over F by $1, a, a^2, \ldots, a^{n-1}$, we have that $[F[a]: F] \le n$. To show that $[F[a]: F] = n$ we must merely show that $1, a, a^2, \ldots, a^{n-1}$ are linearly independent over F. But if $\alpha_0 + \alpha_1 a + \cdots + \alpha_{n-1} a^{n-1} = 0$, with the α_i in F, then $q(a) = 0$, where $q(x) = \alpha_0 + \alpha_1 x + \cdots + \alpha_{n-1} x^{n-1}$ is in $F[x]$. Since $q(x)$ is of lower degree than $p(x)$, which is the minimal polynomial for a in $F[x]$, we are forced to conclude that $q(x) = 0$. This implies that $\alpha_0 = \alpha_1 = \cdots$ $= \alpha_{n-1} = 0$. Therefore, $1, a, a^2, \ldots, a^{n-1}$ are linearly independent over F and form a basis of $F[a]$ over F. Thus $[F[a]: F] = n$. Since $F[a]$ is a field, not merely just a set of polynomial expressions in a, *we shall denote* $F[a]$ *by* $F(a)$. Note also that if M is any field that contains both F and a, then M contains all polynomial expressions in a over F, hence $M \supset F(a)$. So $F(a)$ *is the smallest subfield of K containing both F and a.*

Definition. $F(a)$ is called the field or extension obtained by *adjoining a to F.*

We now summarize.

Theorem 5.3.5. Let $K \supset F$ and suppose that a in K is algebraic over F of degree n. Then $F(a)$, the field obtained by adjoining a to F, is a finite extension of F, and

$$[F(a): F] = n.$$

Before leaving Theorem 5.3.5, let's look at it in a slightly different way. Let $F[x]$ be the polynomial ring in x over F, and let $M = (p(x))$ be the ideal of $F[x]$ generated by $p(x)$, the minimal polynomial for a in K over F. By Lemma 5.3.4, $p(x)$ is irreducible in $F[x]$; hence, by Theorem 4.5.11, M is a maximal ideal of $F[x]$. Therefore, $F[x]/(p(x))$ is a field by Theorem 4.4.2.

Define the mapping $\psi: F[x] \to K$ by $\psi(f(x)) = f(a)$. The mapping ψ is a homomorphism of $F[x]$ into K, and the image of $F[x]$ in K is merely $F(a)$ by the definition of $F(a)$. What is the kernel of ψ? It is by definition $J = \{f(x) \in F[x] | \psi(f(x)) = 0\}$, and since we know

$\psi(f(x)) = f(a)$, $J = \{f(x) \in K | f(a) = 0\}$. Since $p(x)$ is in J and $p(x)$ is the minimal polynomial for a over F, $p(x)$ is of the lowest possible degree among the elements of J. Thus $J = (p(x))$ by the proof of Theorem 4.5.6, and so $J = M$. By the First Homomorphism Theorem for rings $F[x]/M \simeq$ image of $F[x]$ under $\psi = F(a)$, and since $F[x]/M$ is a field, we have that $F(a)$ is a field. We leave the proof, from this point of view, of $[F(a) : F] = \deg p(x)$ to the reader.

Problems

1. Show that the following numbers in $\mathbb{C}$ are algebraic numbers.
   (a) $\sqrt{2} + \sqrt{3}$.
   (b) $\sqrt{7} + \sqrt[3]{12}$.
   (c) $2 + i\sqrt{3}$.
   (d) $\cos(2\pi/k) + i\sin(2\pi/k)$, k a positive integer.

2. Determine the degrees over $\mathbb{Q}$ of the numbers given in Parts (a) and (c) of Problem 1.

3. What is the degree of $\cos(2\pi/3) + i\sin(2\pi/3)$ over $\mathbb{Q}$?

4. What is the degree of $\cos(2\pi/8) + i\sin(2\pi/8)$ over $\mathbb{Q}$?

5. If p is a prime number, prove that the degree of $\cos(2\pi/p) + i\sin(2\pi/p)$ over $\mathbb{Q}$ is $p - 1$ and that

$$f(x) = 1 + x + x^2 + \cdots + x^{p-1}$$

   is its minimal polynomial over $\mathbb{Q}$.

6. (For those who have had calculus) Show that

$$e = 1 + \frac{1}{1!} + \frac{1}{2!} + \cdots + \frac{1}{n!} + \cdots$$

   is irrational.

7. If a in K is such that a^2 is algebraic over the subfield F of K, show that a is algebraic over F.

8. If $F \subset K$ and $f(a)$ is algebraic over F, where $f(x)$ is of positive degree in $F[x]$ and $a \in K$, prove that a is algebraic over F.

9. In the discussion following Theorem 5.3.5, show that $F[x]/M$ is of degree $n = \deg p(x)$ over F, and so $[F(a): F] = n = \deg p(x)$.

10. Prove that $\cos 1°$ is algebraic over $\mathbb{Q}$. ($1°$ = one degree.)

11. If $a \in K$ is transcendental over F, let $F(a) = \{ f(a)/g(a) | f(x), g(x) \neq 0 \in F[x]\}$. Show that $F(a)$ is a field and is the smallest subfield of K containing both F and a.

12. If a is as in Problem 11, show that $F(a) \simeq F(x)$, where $F(x)$ is the field of rational functions in x over F.

13. Let K be a finite field and F a subfield of K. If $[K: F] = n$ and F has q elements, show that K has q^n elements.

14. Using the result of Problem 13, show that a finite field has p^n elements for some prime p and some positive integer n.

15. Construct two fields K and F such that K is an algebraic extension of F but is not a finite extension of F.

4. FINITE EXTENSIONS

We continue in the vein of the preceding section. Again $K \supset F$ will always denote two fields, and we shall use Roman letters for elements of K and Greek ones for those of F.

Let $E(K)$ be the set of all elements in K that are algebraic over F. Certainly, $F \subset E(K)$. Our objective is to prove that $E(K)$ is a field. Once this is done, we'll see a little of how $E(K)$ sits in K.

Without further ado we proceed to

Theorem 5.4.1. $E(K)$ is a subfield of K.

Proof.   What we must show is that if $a, b \in K$ are algebraic over F, then $a \pm b$, ab, and a/b (if $b \neq 0$) are all algebraic over F. This will assure us that $E(K)$ is a subfield of K. We'll do all of $a \pm b$, ab, and a/b in one shot.

Let $K_0 = F(a)$ be the subfield of K obtained by adjoining a to F. Since a is algebraic over F, say of degree m, then, by Theorem 5.3.5, $[K_0: F] = m$. Since b is algebraic over F and since K_0 contains F, we certainly have that b is algebraic over K_0. If b is algebraic over F of degree n, then it is algebraic over K_0 of degree at most n. Thus $K_1 = K_0(b)$, the subfield of K obtained by adjoining b to K_0, is a finite extension of K_0 and $[K_1 : K_0] \leq n$.

Thus, by Theorem 5.3.1, $[K_1 : F] = [K_1 : K_0][K_0 : F] \leq mn$; that is, K_1 is a finite extension of F. As such, by Theorem 5.3.2, K_1 is an algebraic extension of F, so all its elements are algebraic over F. Since $a \in K_0 \subset K_1$ and $b \in K_1$, then all of the elements $a \pm b$, ab, a/b are in K_1, hence are algebraic over F. This is exactly what we wanted. The theorem is proved. $\square$

If we look at the proof a little more carefully, we see that we have actually proved a little more, namely the

> **Corollary.** If a and b in K are algebraic over F of degrees m and n, respectively, then $a \pm b$, ab, and a/b (if $b \neq 0$) are algebraic over F of degree at most mn.

A special case, but one worth noting and recording, is the case $K = \mathbb{C}$ and $F = \mathbb{Q}$. In that case we called the algebraic elements in $\mathbb{C}$ over $\mathbb{Q}$ the *algebraic numbers*. So theorem 5.4.1 in this case becomes

> **Theorem 5.4.2.** The algebraic numbers form a subfield of $\mathbb{C}$.

For all we know at the moment, the set of algebraic numbers may very well be all of $\mathbb{C}$. This is not the case, for transcendental numbers do exist; we show this to be true in Section 6 of Chapter 6.

We return to a general field K. Its subfield $E(K)$ has a very particular quality, which we prove next. This property is that any element in K which is algebraic over $E(K)$ must already be in $E(K)$.

In order not to digress in the course of the proof we are about to give, we introduce the following notation. If $a_1, a_2, \ldots, a_n$ are in K, then $F(a_1, \ldots, a_n)$ will be the field obtained as follows: $K_1 = F(a_1)$, $K_2 = K_1(a_2) = F(a_1, a_2)$, $K_3 = K_2(a_3) = F(a_1, a_2, a_3), \ldots, K_n = K_{n-1}(a_n) = F(a_1, a_2, \ldots, a_n)$.

We now prove

> **Theorem 5.4.3.** If u in K is algebraic over $E(K)$, then u is in $E(K)$.

Proof. To prove the theorem, all we must do is show that u is algebraic over F; this will put u in $E(K)$, and we will be done.

Since u is algebraic over $E(K)$, there is a nontrivial polynomial $f(x) = x^n + a_1 x^{n-1} + a_2 x^{n-2} + \cdots + a_n$, where $a_1, a_2, \ldots, a_n$ are in $E(K)$, such that $f(u) = 0$. Since $a_1, a_2, \ldots, a_n$ are in $E(K)$, they are algebraic over F of degrees, say, $m_1, m_2, \ldots, m_n$, respectively. We claim that $[F(a_1, \ldots, a_n) : F]$ is at most $m_1 m_2 \cdots m_n$. To see this, merely carry out n successive applications of Theorem 5.3.1 to the sequence $K_1, K_2, \ldots, K_n$ of fields defined above. We leave its proof to the reader. Thus, since u is algebraic over the field $K_n = F(a_1, a_2, \ldots, a_n)$ [after all, the polynomial satisfied by u is $p(x) = x^n + a_1 x^{n-1} + \cdots + a_n$, which has all its coefficients in $F(a_1, a_2, \ldots, a_n)$], the field $K_n(u)$ is a finite extension of K_n, and since K_n is a finite extension of F, we have, again by Theorem 5.3.1, that $K_n(u)$ is a finite extension of F. Because $u \in K_n(u)$, we obtain from Theorem 5.3.2 that u is algebraic over F. This puts u in $E(K)$ by the very definition of $E(K)$, thereby proving the theorem. $\square$

There is a famous theorem due to Gauss, often referred to as the *Fundamental Theorem of Algebra*, which asserts (in terms of extension) that the only finite extension of $\mathbb{C}$, the field of complex numbers, is $\mathbb{C}$ itself. In reality this result is not a purely algebraic one, its validity depending heavily on topological properties of the field of real numbers. Be that as it may, it is an extremely important theorem in algebra and in many other parts of mathematics.

The formulation of the Fundamental Theorem of Algebra in terms of the nonexistence of finite extensions of $\mathbb{C}$ is a little different from that which is usually given. The most frequent form in which this famous result is stated involves the concept of a root of a polynomial, a concept we shall discuss at some length later. In these terms the Fundamental Theorem of Algebra becomes: A polynomial of positive degree having coefficients in $\mathbb{C}$ has at least one root in $\mathbb{C}$. The exact meaning of this statement and its equivalence with the other form of the theorem stated above will become clearer later, after the development of the material on roots.

A field L with the property of $\mathbb{C}$ described in the paragraphs above is said to be *algebraically closed*. If we grant that $\mathbb{C}$ is algebraically closed (Gauss' Theorem), then, by Theorem 5.4.3, we also have

The field of algebraic numbers is algebraically closed.

Problems

1. Show that $a = \sqrt{2} - \sqrt{3}$ is algebraic over $\mathbb{Q}$ of degree at most 4 by exhibiting a polynomial $f(x)$ of degree 4 over $\mathbb{Q}$ such that $f(a) = 0$.

2. If a and b in K are algebraic over F of degrees m and n, respectively, and if m and n are relatively prime, show that $[F(a, b) : F] = mn$.

3. If $a \in \mathbb{C}$ is such that $p(a) = 0$, where

$$p(x) = x^5 + \sqrt{2}\,x^3 + \sqrt{5}\,x^2 + \sqrt{7}\,x + \sqrt{11},$$

   show that a is algebraic over $\mathbb{Q}$ of degree at most 80.

4. If $K \supset F$ is such that $[K : F] = p$, p a prime, show that $K = F(a)$ for *every* a in K that is not in F.

5. If $[K : F] = 2^n$ and T is a subfield of K, show that $[T : F] = 2^m$ for some $m \le n$.

6. Give an example of two algebraic numbers a and b of degrees 2 and 3, respectively, such that ab is of degree *less* than 6 over $\mathbb{Q}$.

7. If $K \supset F$ are fields and $a_1, \ldots, a_n$ are in K, show that $F(a_1, \ldots, a_n)$ equals $F(a_{\sigma(1)}, \ldots, a_{\sigma(n)})$ for any permutation σ of $1, 2, \ldots, n$.

5.   CONSTRUCTIBILITY

In ancient Greece, unlike in the other cultures of the time, the Greek mathematicians were interested in mathematics as an abstract discipline rather than as a pragmatic bag of tricks to do accounts or to carry out measurements. They developed strong interests and results in number theory and, most especially, in geometry. In these areas they posed penetrating questions. The questions they asked in geometry—two of which will make up the topic treated here—are still of interest and substance. The English mathematician G. H. Hardy, in his sad but charming little book *A Mathematician's Apology*, describes the ancient Greek mathematicians as "colleagues from another college."

Two of these Greek questions will be our concern in this section. But, as a matter of fact, the answer to both will emerge as a consequence of the criterion for constructibility, which we will obtain. We state these questions now and will explain a little later what is entailed in them.

QUESTION 1.   Can one duplicate a cube using just straight-edge and compass? (By duplicating a cube, we mean doubling its volume).

QUESTION 2.   Can one trisect an arbitrary angle using just straight-edge and compass?

Despite the seemingly infinite number of angle-trisectors that crop up every year, the answer to both questions is "no." As we shall see, it is impossible to trisect 60° using just straight-edge and compass. Of course, some angles are trisectable, for instance, 0°, 90°, 145°, 180°, ..., but most angles (in a very precise meaning of "most") are not.

Before getting to the exact meaning of the questions themselves, we want to spell out in explicit terms exactly what the rules of the game are. *By a straight-edge we do not mean a ruler*—that is, an instrument for measuring arbitrary lengths. *No!* A straight-edge is merely a straight line, with no quantitative or metric properties attributed to it. We are given a line segment—to which we assign length 1—and all other lengths that we get from this must be obtainable merely employing a straight-edge and compass.

Let us call a nonnegative real number, b, a *constructible length* if, by a *finite* number of applications of the straight-edge and compass and the points of intersection obtained between lines and circles so constructed, we can construct a line segment of length b, starting out from the line segment we have assigned length 1.

From our high school geometry we recall some things we can do in this framework.

1. Whatever length we construct on one line can be constructed on any other line by use of the compass acting as a transfer agent.
2. We can draw a line parallel to a given line that goes through a given point.
3. We can construct a length n for any nonnegative integer n.

From these and by using results about the similarity of triangles, we can construct any nonnegative rational length. We don't do that at this moment for it will come out as a special case of what we are about to do.

We claim the following properties:

1. If a and b are constructible lengths, then so is $a + b$. For if AB is a line segment of length a and CD is one of length b, we can transfer this line segment CD, by means of a compass, to obtain the line ABE, where AB is of length a and BE is of length b. Thus the line segment AE is of length $a + b$. If $b > a$, how would you construct $b - a$?

2. If a and b are constructible lengths, then so is ab. We may assume that $a \neq 0$ and $b \neq 0$, otherwise, the statement is trivial. Consider the diagram

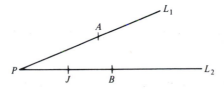

Where L_1 and L_2 are two distinct lines intersecting at P, and such that PA has length a, PB has length b, and PJ has length 1. Let L_3 be the straight line through J and A and L_4 the line parallel to L_3 passing through B. If C is the point of intersection of L_1 and L_4, we have the diagram

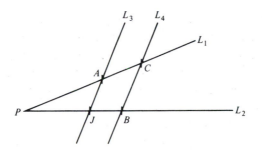

All of these constructions can be carried out by straight-edge and compass. From elementary geometry the length of PC is ab. Therefore, ab is constructible.

3. If a and b are constructible and $b \neq 0$, then a/b is constructible. Consider the diagram

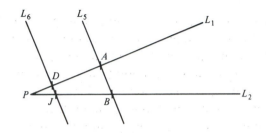

where P, A, B, J, L_1, and L_2 are as in Property 2 above. Let L_5 be the line through A and B and let L_6 be the line through J parallel to L_5. If D is the point of intersection of L_1 and L_6, then, again by elementary geometry, the length of PD is a/b. We stress again that all the constructions made can be carried out by straight-edge and compass.

Of course, this shows that the nonnegative rational numbers are constructible lengths, since they are quotients of nonnegative integers, which we know to be constructible lengths. But there are other constructible lengths, for instance, the irrational number $\sqrt{2}$. Because we can construct by straight-edge and compass the right-angle triangle

with sides AB and BC of length 1, we know, by the Pythagorean Theorem, that AC is of length $\sqrt{2}$. So $\sqrt{2}$ is a constructible length.

In Properties 1 to 3 we showed that the constructible lengths almost form a field. What is lacking is the negatives. To get around this, we make the

> **Definition.**   The real number a is said to be a *constructible number* if $|a|$, the absolute value of a, is a constructible length.

As far as we can say at the moment, any real number might be a constructible one. We shall soon have a criterion which will tell us that certain real numbers are not constructible. For instance, we shall be able to deduce from this criterion that both $\sqrt[3]{2}$ and $\cos 20°$ are *not* constructible. This in turn will allow us to show that the answer to both Questions 1 and 2 is "no."

But first we state

Theorem 5.5.1. The constructible numbers form a subfield of the field of real numbers.

Proof. Properties 1 to 3 almost do the trick; we must adapt Property 1 slightly to allow for negatives. We leave the few details to the reader. □

Our next goal is to show that a constructible number must be an algebraic number—not any old algebraic number, but one satisfying a rather stringent condition.

Note, first, that if $a \geq 0$ is a constructible number, then so is $\sqrt{a}$. Consider the diagram

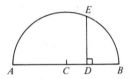

It is of a semicircle of radius $(a + 1)/2$, center at C, AD is of length a, DB is of length 1, and DE is the perpendicular to AB at D, intersecting the circle at E. All this is constructible by straight-edge and compass. From elementary geometry we have that DE is of length $\sqrt{a}$, hence $\sqrt{a}$ is constructible.

We now head for the necessary condition that a real number be constructible. Let K be the field of constructible numbers, and let K_0 be a subfield of K. By the *plane of K_0* we shall mean the set of all points (a, b) in the real Euclidean plane whose coordinates a and b are in K_0. If (a, b) and (c, d) are in the plane of K_0, then the straight line joining them has the equation $(y - b)/(x - a) = (b - d)/(a - c)$, so is of the form $ux + vy + w = 0$, where u, v, and w are in K_0. Given two such lines $u_1x + v_1y + w_1 = 0$ and $u_2x + v_2y + w_2 = 0$, where u_1, v_1, w_1 and u_2, v_2, w_2 are all in K_0, either they are parallel or their point of intersection is a point in K_0. (Prove!)

Given a circle whose radius r is in K_0 and whose center (a, b) is in the plane of K_0, then its equation is $(x - a)^2 + (y - b)^2 = r^2$, which we see, on expanding, is of the form $x^2 + y^2 + dx + ey + f = 0$, where d, e, and f are in K_0. To see where this circle intersects a line in the plane of K_0, $ux + vy + w = 0$, we solve simultaneously between the equation of the line and of the circle. For instance, if $v \neq 0$, then $y = -(ux + w)/v$; substituting this for y in the equation of the circle $x^2 + y^2 + dx + ey + f = 0$ leads us to a quadratic equation for the x-coordinate, c, of this intersection point, of the form $c^2 + s_1 c + s_2 = 0$, with s_1 and s_2 in K_0. By the quadratic formula, $c = (-s_1 \pm \sqrt{s_1^2 - 4s_2})/2$, and if the line and circle intersect in the real plane, then $s_1^2 - 4s_2 \geq 0$. If $s = s_1^2 - 4s_2 \geq 0$ and if $K_1 = K_0(\sqrt{s})$, then we see that the x-coordinate, c, lies in K_1. If $\sqrt{s} \in K_0$, then $K_1 = K_0$; otherwise, $[K_1 : K_0] = 2$. Since the y-coordinate $d = (-uc + w)/v$, we have that d is also in K_1. Thus the intersection point (c, d) lies in the plane of K_1 where $[K_1 : K_0] = 1$ or 2. The story is similar if $v = 0$ and $u \neq 0$.

Finally, to get the intersection of two circles $x^2 + y^2 + dx + ey + f = 0$ and $x^2 + y^2 + gx + hy + k = 0$ in the plane of K_0, subtracting one of these equations from the other gives us the equation of the line in the plane of K_0, $(d - g)x + (e - h)y + (f - k) = 0$. So to find the points of two circles in the plane of K_0 is the same as finding the points of intersection of a line in the plane of K_0 with a circle in that plane. This is precisely the situation we disposed of above. So if the two circles intersect in the real plane, their points of intersection lie in the plane of an extension of K_0 of degree 1 or 2.

To construct a constructible length, a, we start in the plane of $\mathbb{Q}$, the rationals; the straight-edge gives us lines in the plane of $\mathbb{Q}$, and the compass circles in the plane of $\mathbb{Q}$. By the above, these intersect at a point in the plane of an extension of degree 1 or 2 of $\mathbb{Q}$. To get to a, we go by this procedure from the plane of $\mathbb{Q}$ to that of L_1, say, where $[L_1 : \mathbb{Q}] = 1$ or 2, then to that of L_2, where $[L_2 : L_1] = 1$ or 2, and continue a finite number of times. We get, this way, a finite sequence $\mathbb{Q} = L_0 \subset L_1 \subset \cdots \subset L_n$ of fields, where each $[L_i : L_{i-1}] = 1$ or 2 and where a is in L_n.

By Theorem 5.3.1, $[L_n : \mathbb{Q}] = [L_n : L_{n-1}][L_{n-1} : L_{n-2}] \cdots [L_1 : \mathbb{Q}]$ and since each of $[L_i : L_{i-1}] = 1$ or 2, we see that $[L_n : \mathbb{Q}]$ is a *power of 2*. Since $a \in L_n$, we have that $\mathbb{Q}(a)$ is a subfield of L_n, hence by the *Corollary to Theorem 5.3.1*, $[\mathbb{Q}(a) : \mathbb{Q}]$ *must divide a power of 2, hence*

$[\mathbb{Q}(a):\mathbb{Q}] = 2^m$ *for some nonnegative integer m*. Equivalently, by Theorem 5.3.5, the minimal polynomial for a over $\mathbb{Q}$ must have degree a power of 2. This is a *necessary* condition that a be constructible. We have proved the important criterion for constructibility, namely

> **Theorem 5.5.2.** In order that the real number a be constructible, it is necessary that $[\mathbb{Q}(a):\mathbb{Q}]$ be a power of 2. Equivalently, the minimal polynomial of a over $\mathbb{Q}$ must have degree a power of 2.

To duplicate a cube of sides 1, so of volume 1, by straight-edge and compass would require us to construct a cube of sides of length b whose volume would be 2. But the volume of this cube would be b^3, so we would have to be able find a constructible number b such that $b^3 = 2$.

Given a real number b such that $b^3 = 2$, then its minimal polynomial over $\mathbb{Q}$ is $p(x) = x^3 - 2$, for this polynomial is monic and irreducible over $\mathbb{Q}$ (if you want, by the Eisenstein Criterion), and $p(b) = 0$. Also, as is clear to the eye, $p(x)$ is of degree 3. Since 3 is not a power of 2, by Theorem 5.5.2, there is no such constructible b. Therefore, the question of the duplication of the cube by straight-edge and compass has a negative answer. We summarize this in

> **Theorem 5.5.3.** It is impossible to duplicate a cube of volume 1 by straight-edge and compass.

We now have disposed of the classical Question 1, so we turn our attention to Question 2, the trisection of an arbitrary angle by straight-edge and compass.

If we could trisect the particular angle 60°, we would be able to construct the triangle ABC in the diagram where $\theta = 20°$ and AC is of length 1, by straight-edge and compass. Since AB is of length $\cos 20°$, we would have that $b = \cos 20°$ is a constructible number.

We want to show that $b = \cos 20°$ is not a constructible number by producing its minimal polynomial over $\mathbf{Q}$, and showing that this polynomial is of degree 3. To this end we recall the triple-angle formula from trigonometry, namely that $\cos 3\phi = 4\cos^3 \phi - 3\cos \phi$. If $b = \cos 20°$, then, since $\cos(3 \cdot 20°) = \cos 60° = \frac{1}{2}$, this trigonometric formula becomes $4b^3 - 3b = \frac{1}{2}$, and so $8b^3 - 6b - 1 = 0$. If $c = 2b$, this becomes $c^3 - 3c - 1 = 0$. If b is constructible, then so is c. But $p(c) = 0$, where $p(x) = x^3 - 3x - 1$, and this polynomial is irreducible over $\mathbf{Q}$. (Prove!) So $p(x)$ is the minimal polynomial for c over $\mathbf{Q}$. Because $p(x)$ is of degree 3, and 3 is not a power of 2, by Theorem 5.5.1 we have that c is not constructible. So we *cannot* trisect $60°$ by straight-edge and compass. This answers Question 2 in the negative.

Theorem 5.5.4. It is impossible to trisect $60°$ by straight-edge and compass.

We hope that this theorem will dissuade any reader from joining the hordes of angle-trisectors. There are more profitable and pleasanter ways of wasting one's time.

There is yet another classical problem of this kind to which the answer is "no." This is the question of squaring the circle. This question asks: Can we construct a square whose area is that of a circle of radius 1 by straight-edge and compass? This is equivalent to asking whether $\sqrt{\pi}$ is a constructible number. If this were the case, then since $\pi = (\sqrt{\pi})^2$, the number π would be constructible. But Lindemann proved in 1882 that π is in fact transcendental, so certainly is not algebraic, and so cannot be constructible. Therefore, the circle of radius 1 cannot be squared by straight-edge and compass.

Of course, what we did above does not constitute a proof of the impossibility of squaring the circle, since we have presupposed Lindemann's result without proving it. To prove that π is transcendental would take us too far afield. One might expect that it would be easier to prove that π is not constructible than to prove that it is not algebraic. This does not seem to be the case. Until now all proofs that π is not constructible go via the route of exploiting the transcendence of π.

Problems

1. Complete the proof of Theorem 5.5.1.

2. Prove that $x^3 - 3x - 1$ is irreducible over $\mathbb{Q}$.

3. Show that the construction given for $\sqrt{a}$, $a \geq 0$ does indeed give us $\sqrt{a}$.

4. Prove that the regular heptagon (seven-sided polygon with sides of equal length) is not constructible by straight-edge and compass.

6. ROOTS OF POLYNOMIALS

Let $F[x]$, as usual, be the polynomial ring in x over the field F and let K be an extension field of F. If $a \in K$ and

$$f(x) = \alpha_0 + \alpha_1 x + \cdots + \alpha_n x^n,$$

then by $f(a)$ we understand the element

$$f(a) = \alpha_0 + \alpha_1 a + \cdots + \alpha_n a^n$$

in K. This is the usage we have made of this notation throughout this chapter. We will now be interested in those a's in K such that $f(a) = 0$.

Definition. The element $a \in K$ is a *root* of the polynomial $f(x) \in F[x]$ if $f(a) = 0$.

In what we have done up until now we have always had an extension field K of F given to us and we considered the elements in K algebraic over F, that is, those elements of K that are roots of nonzero polynomials in $F[x]$. We saw that if $a \in K$ is algebraic over F of degree n—that is, if the minimal polynomial for a over F is of degree n—then $[F(a): F] = n$, where $F(a)$ is the subfield of K obtained by adjoining a to F.

What we do now is turn the problem around. We no longer will have the extension K of F at our disposal. In fact, our principal task will be to produce it almost from scratch. We start with some polynomial $f(x)$ of positive degree in $F[x]$ as our only bit of data; our goal is to construct an extension field K of F in which $f(x)$ will have a root. Once we have this construction of K under control, we shall elaborate on the general theme, thereby obtaining a series of interesting consequences.

Before setting off on this search for the appropriate K, we must get some information about the relation of the roots of a given polynomial and the factorization of that polynomial.

> **Lemma 5.6.1.** If $a \in L$ is a root of the polynomial $f(x) \in F[x]$ of degree n, where L is an extension field of F, then $f(x)$ factors in $L[x]$ as $f(x) = (x - a)q(x)$, where $q(x)$ is of degree $n - 1$ in $L[x]$. Conversely, if $f(x) = (x - a)q(x)$, with $f(x)$, $q(x)$, and a as above, then a is a root of $f(x)$ in L.

Proof. Since $F \subset L$, $F[x]$ is contained in $L[x]$. Because $a \in L$, $x - a$ is in $L[x]$; by the Division Algorithm, Theorem 4.5.5, for polynomials in $L[x]$ we have that $f(x) = (x - a)q(x) + r(x)$, where $q(x)$ and $r(x)$ are in $L[x]$ and where $r(x) = 0$ or $\deg r(x) < \deg(x - a) = 1$. This yields that $r(x) = b$, some element of L. Substituting a for x in the relation above, and using that $f(a) = 0$, we obtain $0 = (a - a)q(a) + b = 0 + b = b$; thus $b = 0$. Since $r(x) = b = 0$, we have what we wanted, namely $f(x) = (x - a)q(x)$.

For the statement that $\deg q(x) = n - 1$ we note that since $f(x) = (x - a)q(x)$, then, by Lemma 4.5.2, $n = \deg f(x) = \deg(x - a) + \deg q(x) = 1 + \deg q(x)$. This gives us the required result, $\deg q(x) = n - 1$.

The converse is completely trivial. $\square$

One immediate consequence of Lemma 5.6.1 is

> **Theorem 5.6.2.** Let $f(x)$ in $F[x]$ have degree n; then $f(x)$ can have at most n roots in any extension, K, of F.

Proof. We go by induction on n. If $n = 1$, then $f(x) = ax + b$, where a and b are in F and where $a \neq 0$. Thus the only root of $f(x)$ is $-b/a$, an element of F.

Suppose that the theorem is correct for all polynomials of degree $k - 1$ over *any* field. Suppose that $f(x)$ in $F[x]$ is of degree k. If $f(x)$ has no roots in K, then the theorem is certainly correct. Suppose, then, that $a \in K$ is a root of $f(x)$. By Lemma 5.6.1, $f(x) = (x - a)q(x)$, where $q(x)$ is of degree $k - 1$ in $K[x]$. Any root b in K of $q(x)$ is either a or is a root of $q(x)$, since $0 = f(b) = (b - a)q(b)$. By induction, $q(x)$ has at most $k - 1$ roots in K, hence $f(x)$ has at most k roots in K. This completes the induction and proves the theorem. □

Actually, the proof yields a little more. To explain this "little more," we need the notion of the multiplicity of a root.

Definition. If K is an extension of F, then the element a in K is a *root of multiplicity $k > 0$ of $f(x)$*, where $f(x)$ is in $F[x]$, if $f(x) = (x - a)^k q(x)$ for some $q(x)$ in $K[x]$ and $x - a$ does *not* divide $q(x)$ (or, equivalently, where $q(a) \neq 0$).

The same proof as that given for Theorem 5.6.2 yields the sharpened version:

Let $f(x)$ be a polynomial of degree n in $F[x]$; then $f(x)$ can have at most n roots in any extension field K of F, counting a root of multiplicity k as k roots.

Theorem 5.6.3. Let $f(x)$ in $F[x]$ be monic of degree n and suppose that K is an extension of F in which $f(x)$ has n roots, counting a root of multiplicity k as k roots. If these roots in K are $a_1, a_2, \ldots, a_m$, each having multiplicity $k_1, k_2, \ldots, k_m$ respectively, then $f(x)$ factors in $K[x]$ as $f(x) = (x - a_1)^{k_1}(x - a_2)^{k_2} \cdots (x - a_m)^{k_m}$.

Proof. The proof is easy by making use Lemma 5.6.1 and of induction on n. We leave the carrying out of the proof to the reader. □

Definition. We say that $f(x)$ in $F[x]$ *splits into linear factors over (or, in) K* if $f(x)$ has the factorization in $K[x]$ given in Theorem 5.6.3.

There is a nice application of Theorem 5.6.3 to finite fields. Let F be a finite field having q elements, and let $a_1, a_2, \ldots, a_{q-1}$ be the nonzero elements of F. Since these form a group of order $q - 1$ under the multiplication in F, by Theorem 2.4.5 (proved ever so long ago), $a^{q-1} = 1$ for any $a \neq 0$ in F. Thus the polynomial $x^{q-1} - 1$ in $F[x]$ has $q - 1$ distinct roots in F. By Theorem 5.6.3, the polynomial $x^{q-1} - 1 = (x - a_1)(x - a_2) \cdots (x - a_{q-1})$. If we also consider 0, then every element a in F satisfies $a^q = a$, so that the polynomial $x^q - x$ has the q elements of F as its distinct roots. By Theorem 5.6.3 we have

Theorem 5.6.4. Let F be a finite field having q elements. Then $x^q - x$ factors in $F[x]$ as

$$x^q - x = x(x - a_1)x - a_2) \cdots (x - a_{q-1}),$$

where $a_1, a_2, \ldots, a_{q-1}$ are the nonzero elements of F, and

$$x^{q-1} - 1 = (x - a_1)(x - a_2) \cdots (x - a_{q-1}).$$

A very special case of this theorem is that in which $F = \mathbb{Z}_p$, the integers modulo the prime p. Here $q = p$ and $a_1, a_2, \ldots, a_{p-1}$ are just $1, 2, \ldots, p - 1$ in some order. Thus the

Corollary. In $\mathbb{Z}_p[x]$, $x^{p-1} - 1$ factors as

$$x^{p-1} - 1 = (x - 1)(x - 2) \cdots (x - (p - 1)).$$

Try this out for $p = 5$, 7, and 11.

As a corollary to the corollary, we have a result in number theory, known as *Wilson's Theorem*, which we assigned as Problem 18 in Section 4 of Chapter 2.

Corollary. If p is a prime, then $(p - 1)! \equiv -1 \bmod p$.

Proof. By the Corollary above,

$$x^{p-1} - 1 = (x - 1)(x - 2) \cdots (x - (p - 1));$$

substituting $x = 0$ in this gives us

$$-1 = (-1)(-2) \cdots (-(p-1)) = (-1)^{p-1} 1 \cdot 2 \cdots (p-1)$$

$$= (-1)^{p-1}(p-1)!$$

in Z_p. In the integers this translates into "congruent mod p." Thus

$$(-1)^{p-1}(p-1)! \equiv -1 \bmod p,$$

and so $(p-1)! \equiv (-1)^p \bmod p$. But $(-1)^p \equiv -1 \bmod p$; hence we have proved Wilson's Theorem. $\square$

We change direction to consider the problem mentioned at the beginning of this section: given $f(x) \in F[x]$, to construct a finite extension of K of F in which $f(x)$ has a root. As we shall see in a moment, this construction of K will be quite easy when we bring the results about polynomial rings proved in Chapter 4 into play. However, to verify that this construction works will take a bit of work.

> **Theorem 5.6.5.** Let F be a field and $f(x)$ a polynomial of positive degree n in $F[x]$. Then there exists a finite extension K of F, with $[K : F] \leq n$, in which $f(x)$ has a root.

Proof. By Theorem 4.5.12, $f(x)$ is divisible in $F[x]$ by some irreducible polynomial $p(x)$ in $F[x]$. Since $p(x)$ divides $f(x), \deg p(x) \leq \deg f(x) = n$, and $f(x) = p(x)q(x)$ for some polynomial $q(x)$ in $F[x]$. If b is a root of $p(x)$ in some extension field, then b is automatically a root of $f(x)$, since $f(b) = p(b)q(b) = 0q(b) = 0$. So to prove the theorem it is enough to find an extension of F in which $p(x)$ has a root.

Because $p(x)$ is irreducible in $F[x]$, the ideal $M = (p(x))$ of $F[x]$ generated by $p(x)$ is a maximal ideal of $F[x]$ by Theorem 4.5.11. Thus by Theorem 4.4.2, $K = F[x]/M$ is a field. We claim that this is the field that we are seeking.

Strictly speaking, K does not contain F; as we now show, however, K does contain a field isomorphic to F. Since every element in M is a multiple in $F[x]$ of $p(x)$, every such nonzero element must have degree at least that of $p(x)$. Therefore, $M \cap F = (0)$. Thus the homomorphism

$\psi : F[x] \to K$ defined by $\psi(g(x)) = g(x) + M$ for every $g(x)$ in $F[x]$, when restricted to F, is $1 - 1$ on F. Therefore, the image $\bar{F}$ of F in K is a field isomorphic to F. We can identify $\bar{F}$, via ψ, with F and so, in this way, we can consider K an extension of F.

Denote $x + M \in K$ by a, so that $\psi(x) = a$, $a \in K$. We leave it to the reader to show, from the fact that ψ is a homomorphism of $F[x]$ onto K with kernel M, that $\psi(g(x)) = g(a)$ for every $g(x)$ in $F[x]$. What is $\psi(p(x))$? On the one hand, since $p(x)$ is in $F[x]$, $\psi(p(x)) = p(a)$. On the other hand, since $p(x)$ is in M, the kernel of ψ, $\psi(p(x)) = 0$. Equating these two evaluations of $\psi(p(x))$, we get that $p(a) = 0$. *In other words, the element $a = \psi(x)$ in K is a root of $p(x)$.*

To finish the proof, all we need is to show that $[K : F] = \deg p(x) \le n$. This came up earlier, in the alternative proof we gave of Theorem 5.3.5. There we left this point to be proved by the reader. We shall be a little more generous here and carry out the proof in detail.

Given $h(x)$ in $F[x]$, then, by the Division Algorithm, $h(x) = p(x)q(x) + r(x)$ where $q(x)$ and $r(x)$ are in $F[x]$, and $r(x) = 0$ or $\deg r(x) < \deg p(x)$. Going modulo M, we obtain that

$$\psi(h(x)) = \psi(p(x)q(x) + r(x)) = \psi(p(x)(q(x)) + \psi(r(x))$$

$$= \psi(p(x))\psi(q(x)) + \psi(r(x))$$

$$= \psi(r(x)) = r(a)$$

[since $\psi(p(x)) = p(a) = 0$].

So, since every element in $K = F[x]/M$ is $\psi(h(x))$ for some $h(x)$ in $F[x]$ and $\psi(h(x)) = r(a)$, we see that every element of K is of the form $r(a)$, where $r(x)$ is in $F[x]$ and $\deg r(x) < \deg p(x)$. If $\deg p(x) = m$, the discussion just made tells us that $1, a, a^2, \ldots, a^{m-1}$ span K over F. Moreover, these elements are linearly independent over F, since a relation of the type $\alpha_0 + \alpha_1 a + \cdots + \alpha_{m-1} a^{m-1} = 0$ would imply that $g(a) = 0$ where $g(x) = \alpha_0 + \alpha_1 x + \cdots + \alpha_{m-1} x^{m-1}$ is in $F[x]$. This puts $g(x)$ in M, which is impossible since $g(x)$ is of lower degree than $p(x)$, unless $g(x) = 0$. In other words, we get a contradiction unless $\alpha_0 = \alpha_1 = \cdots = \alpha_{m-1} = 0$. So the elements $1, a, a^2, \ldots, a^{m-1}$ are linearly independent over F. Since they also span K over F, they form a basis of K over F. Consequently,

$$\dim_F K = [K : F] = m = \deg p(x) \le n = \deg f(x).$$

The theorem is proved. $\square$

We carry out an iteration of the argument used in the last proof to prove the important.

Theorem 5.6.6. Let $f(x) \in F[x]$ be of degree n. Then there exists an extension K of F of degree at most $n!$ over F such that $f(x)$ has n roots, counting multiplicities, in K. Equivalently, $f(x)$ splits into linear factors over K.

Proof. We go by induction on n. If $n = 1$, then $f(x) = \alpha + \beta x$, where $\alpha, \beta \in F$ and where $\beta \ne 0$. The only root of $f(x)$ is $-\alpha/\beta$, which is in F. Thus $K = F$ and $[K : F] = 1$.

Suppose that the result is true for all fields for polynomials of degree k, and suppose that $f(x) \in F[x]$ is of degree $k + 1$. By Theorem 5.6.5 there exists an extension K_1 of F with $[K_1 : F] \le k + 1$ in which $f(x)$ has a root a_1. Thus in $K_1[x]$, $f(x)$ factors as $f(x) = (x - a_1)q(x)$, where $q(x) \in K_1[x]$ is of degree k. By induction there exists an extension K of K_1 of degree at most $k!$ over K_1 over which $q(x)$ splits into linear factors. But then $f(x)$ splits into linear factors over K. Since $[K : F] = [K : K_1][K_1 : K] \le (k + 1)k! = (k + 1)!$, the induction is completed and the theorem is proved. $\square$

We leave the subject of field extensions at this point. We are exactly at what might be described as the beginning of Galois theory. Having an extension K of F of finite degree over which a given polynomial $f(x)$ splits into linear factors, there exists an extension of *least* degree enjoying this property. Such an extension is called a *splitting field* of $f(x)$ over F. One then proceeds to prove that such a splitting field is unique up to isomorphism. Once this is in hand the Galois theory goes into full swing, studying the relationship between the group of automorphisms of this splitting field and its subfield structure. Eventually, it leads to showing, among many other things, that there exist polynomials over the rationals of all degrees 5 or higher whose roots cannot be expressed nicely in terms of the coefficients of these polynomials.

This is a brief and very sketchy description of where we can go from here in field theory. But there is no hurry. The readers should assimilate the material we have presented; this will put them in a good position to learn Galois theory if they are so inclined.

Problems

1. Prove Theorem 5.6.3.

2. If F is a finite field having the $q - 1$ nonzero elements $a_1, a_2, \ldots, a_{q-1}$, prove that $a_1 a_2 \cdots a_{q-1} = (-1)^q$.

3. Let Q be the rational field and let $p(x) = x^4 + x^3 + x^2 + x + 1$. Show that there is an extension K of Q with $[K : Q] = 4$ over which $p(x)$ splits into linear factors. [**Hint:** Find the roots of $p(x)$.]

4. If $q(x) = x^n + a_1 x^{n-1} + \cdots + a_n, a_n \neq 0$, is a polynomial with integer coefficients and if the rational number r is a root of $q(x)$, prove that r is an integer and $r | a_n$.

5. Show that $q(x) = x^3 - 7x + 11$ is irreducible over Q.

6. If F is a field of characteristic $p \neq 0$, show that $(a + b)^p = a^p + b^p$ for all a and b in F.

7. Extend the result of Problem 6 by showing that $(a + b)^m = a^m + b^m$, where $m = p^n$.

8. Let $F = Z_p$, p a prime, and consider the polynomial $x^m - x$ in $Z_p[x]$, where $m = p^n$. Let K be a finite extension of Z_p over which $x^m - x$ splits into linear factors. In K let K_0 be the set of all roots of $x^m - x$. Show that K_0 is a field having at most p^n elements.

9. In Problem 8 show that K_0 has *exactly* p^n elements. (**Hint:** See Problem 14.)

10. Construct an extension field K_n of Q such that $[K_n : Q] = n$, for any $n \geq 1$.

11. Define the mapping $\delta : F[x] \rightarrow F[x]$ by

$$\delta\left(a_0 + a_1 x + a_2 x^2 + \cdots + a_n x^n\right)$$

$$= a_1 + 2a_2 x + \cdots + ia_i x^{i-1} + \cdots + na_n x^{n-1}.$$

Prove that:
(a) $\delta(f(x) + g(x)) = \delta(f(x)) + \delta(g(x))$.
(b) $\delta(f(x)g(x)) = f(x)\delta(g(x)) + \delta(f(x))g(x)$
for all $f(x)$ and $g(x)$ in $F[x]$.

12. If F is of characteristic $p \neq 0$, characterize all $f(x)$ in $F[x]$ such that $\delta(f(x)) = 0$.

13. Show that if $f(x)$ in $F[x]$ has a root of multiplicity greater than 1 in some extension field of F, then $f(x)$ and $\delta(f(x))$ are *not* relatively prime in $F[x]$.

14. If F is of characteristic $p \neq 0$, show that all the roots of $x^m - x$, where $m = p^n$, are distinct.

15. If $f(x)$ in $F[x]$ is irreducible and has a root of multiplicity greater than 1 in some extension of F, show that:
(a) F must be of characteristic p for some prime p.
(b) $f(x) = g(x^p)$ for some polynomial $g(x)$ in $F[x]$.

Chapter 6

Special Topics (Optional)

In this final chapter we treat several unrelated topics. One of these comes from group theory, and all the rest from the theory of fields. In handling these special topics, we draw from many of the results and ideas developed earlier in the book. Although these topics are somewhat special, each of them has results that are truly important in their respective areas.

The readers who have managed to survive so far should have picked up a certain set of techniques, experience, and algebraic know-how to be able to follow the material with a certain degree of ease. We now feel free to treat the various matters at hand in a somewhat sketchier fashion than we have heretofore, leaving a few more details to the reader to fill in.

The material we shall handle does not lend itself readily to problems, at least not to problems of a reasonable degree of difficulty. Accordingly, we will assign relatively few exercises. This should come as a relief to those wanting to assimilate the material in this chapter.

1.   THE SIMPLICITY OF A_n

In Chapter 3, where we discussed S_n, the symmetric group of degree n, we showed that if $n \geq 2$, then S_n has a normal subgroup A_n, which we called the *alternating group of degree n*, which is a group of order $n!/2$. In fact, A_n was merely the set of all even permutations in S_n.

In discussing A_n, we said that A_n, for $n \geq 5$, was a simple group, that is, that A_n has no normal subgroups other than (e) and itself. We promised there that we would prove this fact in Chapter 6. We now make good on this promise.

To make clear what it is that we are about to prove, we should perhaps repeat what we said above and formally define what is meant by a simple group.

Definition.   A nonabelian group is said to be *simple* if its only normal subgroups are (e) and itself.

We impose the proviso that G be nonabelian to exclude the trivial examples of cyclic groups of prime order from the designation "simple." These cyclic groups of prime order have no nontrivial subgroups at all, so, perforce, they have no proper normal subgroups. An abelian group with no proper subgroups is easily seen to be cyclic of prime order.

We begin with the very easy

Lemma 6.1.1.   If $n \geq 3$ and τ_1, τ_2 are two transpositions in S_n, then $\tau_1\tau_2$ is either a 3-cycle or the product of two 3-cycles.

Proof.   If $\tau_1 = \tau_2$, then $\tau_1\tau_2 = \tau_1^2 = e$ and e is certainly the product of two 3-cycles, for instance as $e = (123)(132)$.

If $\tau_1 \neq \tau_2$, then they either have one letter in common or none. If they have one letter in common, we may suppose, on a suitable renumbering, that $\tau_1 = (12)$ and $\tau_2 = (13)$. But then $\tau_1\tau_2 = (12)(13) = (132)$, which is already a 3-cycle.

Finally, if τ_1 and τ_2 have no letter in common, we may suppose, without loss of generality, that $\tau_1 = (12)$ and $\tau_2 = (34)$, in which case

$\tau_1\tau_2 = (12)(34) = (142)(143)$, which is indeed the product of two 3-cycles. The lemma is now proved. $\square$

An immediate consequence of Lemma 6.1.1 is that for $n \geq 3$ the 3-cycles generate A_n, the alternating group of degree n.

Theorem 6.1.2. If σ is an even permutation in S_n, where $n \geq 3$, then σ is a product of 3-cycles. In other words, the 3-cycles in S_n generate A_n.

Proof. Let $\sigma \in S_n$ be an even permutation. By the definition of the parity of a permutation, σ is a product of an even number of transpositions. Thus $\sigma = \tau_1\tau_2 \cdots \tau_{2i-1}\tau_{2i} \cdots \tau_{2m-1}\tau_{2m}$ is a product of $2m$ transpositions $\tau_1, \tau_2, \ldots, \tau_{2m}$. By Lemma 6.1.1, each $\tau_{2i-1}\tau_{2i}$ is either a 3-cycle or a product of two 3-cycles. So we get that σ is either a 3-cycle or the product of at most $2m$ 3-cycles. This proves the theorem. $\square$

We now give an algorithm for computing the conjugate of any permutation in S_n. Let $\sigma \in S_n$, and suppose that $\sigma(i) = j$. What does $\tau\sigma\tau^{-1}$ look like if $\tau \in S_n$? Suppose that $\tau(i) = s$ and $\tau(j) = t$; then $\tau\sigma\tau^{-1}(s) = \tau\sigma(\tau^{-1}(s)) = \tau\sigma(i) = \tau(j) = t$. In other words, to compute $\tau\sigma\tau^{-1}$ replace every symbol in σ by its image under τ.

For instance, if $\sigma = (123)$ and $\tau = (143)$, then, since $\tau(1) = 4$, $\tau(2) = 2$, $\tau(3) = 1$, and $\tau(4) = 3$, we see that $\tau\sigma\tau^{-1} = (421) = (142)$.

Given two k-cycles, say $(12 \cdots k)$ and $(i_1 i_2 \cdots i_k)$, then they are conjugate in S_n because if τ is a permutation that sends 1 into i_1, 2 into $i_2, \ldots, k$ into i_k, then $\tau(12 \cdots k)\tau^{-1} = (i_1 i_2, \ldots, i_k)$. Since every permutation is the product of disjoint cycles and conjugation is an automorphism, we get, from the result for k-cycles, that to compute $\tau\sigma\tau^{-1}$ for *any* permutation σ, replace every symbol in σ by its image under τ. In this way we see that it is extremely easy to compute the conjugate of any permutation.

Given two permutations σ_1 and σ_2 in S_n, then they are conjugate in S_n, using the observation above, if in their decompositions into products of disjoint cycles they have the same cycle lengths and each cycle length with the same multiplicity. Thus, for instance, (12)(34)(567) and (37)(24)(568) are conjugate in S_8, but (12)(34)(567) and (37)(568) are not.

Recall that by a *partition* of the positive integer n, we mean a decomposition of n as $n = n_1 + n_2 + \cdots + n_k$, where $0 \leq n_1 \leq n_2 \leq \cdots \leq n_k$. If σ in S_n is the disjoint product of an n_1-cycle, an n_2-cycle, ..., an n_k-cycle, then $n_1 + n_2 + \cdots + n_k = n$, and a permutation τ is conjugate to σ if and only if τ is the disjoint product of cycles in the same way. Therefore, the number of conjugacy classes in S_n is equal to the number of partitions of n.

For instance, if $n = 4$, then the partitions of 4 are $4 = 4$, $4 = 1 + 3$, $4 = 1 + 1 + 2$, $4 = 1 + 1 + 1 + 1$, and $4 = 2 + 2$, which are five in number. Thus S_4 has five conjugacy classes, namely the classes of (1234), (123), (12), e, and (12)(34), respectively.

We summarize everything we said above into three distinct statements.

Lemma 6.1.3. To find $\tau\sigma\tau^{-1}$ in S_n, replace every symbol in σ by its image under τ.

Lemma 6.1.4. Two elements in S_n are conjugate if they have similar decompositions as the product of disjoint cycles.

Lemma 6.1.5. The number of conjugacy classes in S_n is equal to the number of partitions of n.

Clearly, from the results above, any two 3-cycles in S_n are conjugate in S_n. A 3-cycle is an even permutation, so is in A_n. One might wonder if any two 3-cycles are actually conjugate in the smaller group A_n. For $n \geq 5$ the answer is "yes," and is quite easy to prove.

Lemma 6.1.6. If $n \geq 5$, then any two 3-cycles in S_n are already conjugate in A_n.

Proof. Let σ_1 and σ_2 be two 3-cycles in S_n; by Lemma 6.1.4 they are conjugate in S_n. By renumbering, we may assume that $\sigma_1 = (123)$ and

$\sigma_2 = \tau(123)\tau^{-1}$ for some $\tau \in S_n$. If τ is even, then we are done. If τ is odd, then $\rho = \tau(45)$ is even and $\rho(123)\rho^{-1} = \tau(45)(123)(45)^{-1}\tau^{-1} = \tau(123)\tau^{-1} = \sigma_2$. Therefore, σ_1 and σ_2 are conjugate in A_n. We thus see that the lemma is correct. $\square$

In S_3 the two 3-cycles (123) and (132) are conjugate in S_3 but are not conjugate in A_3, which is cyclic group of order 3.

We now prove a result that is not only important in group theory, but also plays a key role in field theory and the theory of equations.

Theorem 6.1.7. If $n \geq 5$, then the only nontrivial proper normal subgroup of S_n is A_n.

Proof. Suppose that N is a normal subgroup of S_n and N is neither (e) nor S_n. Let $\sigma \neq e$ be in N. Since the center of S_n is just (e) (See Problem 1) and the transpositions generate S_n, there is a transposition τ such that $\sigma\tau \neq \tau\sigma$. By Lemma 6.1.4, $\tau_1 = \sigma\tau\sigma^{-1}$ is a transposition, so $\tau\tau_1 = \tau\sigma\tau\sigma^{-1} \neq e$ is in N, since $\sigma \in N$ and $\tau\sigma\tau = \tau\sigma\tau^{-1} \in N$ because N is normal in S_n. So N contains an element that is the product of two transpositions, namely $\tau\tau_1$.

If τ and τ_1 have a letter in common, then, as we saw in the proof of Lemma 6.1.1, $\tau\tau_1$ is a 3-cycle, hence N contains a 3-cycle. By Lemma 6.1.4 all 3-cycles in S_n are conjugate to $\tau\tau_1$ so must fall in N, by the normality of N in S_n. Thus the subgroup of S_n generated by the 3-cycles, which, according to Theorem 6.1.2, is all of A_n, lies in N. Note that up to this point we have not used that $n \geq 5$.

We may thus assume that τ and τ_1 have no letter in common. Without loss of generality we may assume that $\tau = (12)$ and $\tau_1 = (34)$; therefore, (12)(34) is in N. Since $n \geq 5$, (15) is in S_n hence $(15)(12)(34)(15)^{-1} = (25)(34)$ is also in N; thus (12)(34)(25)(34) = (125) is in N. Thus in this case also, N must contain a 3-cycle. The argument above then shows that $N \supset A_n$.

We have shown that in both cases N must contain A_n. Since there are no subgroups strictly between A_n and S_n and $N \neq S_n$, we obtain the desired result that $N = A_n$. $\square$

The result is false for $n = 4$; the subgroup

$$N = \{ e, (12)(34), (13)(24), (14)(23) \}$$

is a proper normal subgroup of S_4 and is not A_4.

We now know all the normal subgroups of S_n when $n \geq 5$. Can we determine from this all the normal subgroups of A_n for $n \geq 5$? The answer is "yes"; as we shall soon see, A_n is a simple group if $n \geq 5$. The proof we give may strike many as strange, for it hinges on the fact that 60, the order of A_5, is not a perfect square.

Theorem 6.1.8.   The group A_5 is a simple group of order 60.

Proof.   Suppose that A_5 is not simple; then it has a proper normal subgroup N whose *order is as small as possible*. Let the subset $T = \{ \sigma \in S_5 | \sigma N \sigma^{-1} \subset N \}$, the normalizer of N in S_5. Since N is normal in A_5, we know that $T \supset A_5$. T is a subgroup of S_5, so if $T \neq A_5$, we would have that $T = S_5$. But this would tell us that N is normal in S_5, which, by Theorem 6.1.7, would imply that $N \supset A_5$, giving us that $N = A_5$, contrary to our supposition that N is a proper subgroup of A_5. So we must have $T = A_5$. Since (12) is odd, it is not in A_5, hence is not in T. Therefore, $M = (12)N(12)^{-1} \neq N$.

Since $N \triangleleft A_5$, we also have that $M \triangleleft A_5$ (Prove!), thus both $M \cap N$ and $MN = \{ mn | m \in M, n \in N \}$ are normal in A_5. (See Problem 9.) Because $M \neq N$ we have that $M \cap N \neq N$, and since N is a minimal proper normal subgroup of A_5, it follows that $M \cap N = (e)$. On the other hand, $(12) MN (12)^{-1} = (12) M (12)^{-1} (12) N (12)^{-1} = NM$ (since $(12) N (12)^{-1} = M$ and $(12) M (12)^{-1} = N) = MN$ by the normality of M and N in A_5. Therefore, the element (12) is in the normalizer of MN in S_5, and since MN is normal in A_5, we get, as we did above, the MN is normal in S_5, and so $MN = A_5$ by Theorem 6.17.

Consider what we now have. Both M and N are normal subgroups of A_5, each of order $|N|$, and $MN = A_5$ and $M \cap N = (e)$. We claim, and leave to the reader, that MN must then have order $|N|^2$. Since $MN = A_5$, we obtain that $60 = |A_5| = |MN| = |N|^2$. But this is sheer nonsense, since 60 is not the square of any integer. This establishes Theorem 6.1.8. $\square$

To go from the simplicity of A_5 to that of A_n for $n \geq 5$ is not too hard. Note that the argument we gave for A_5 did not depend on 5 until the punch line "60 is not the square of any integer." In fact, the reasoning is valid as long as we know that $n!/2$ is not a perfect square. Thus, for example, if $n = 6$, then $6!/2 = 360$ is not a square, hence A_6 is a simple group. Since we shall need this fact in the subsequent discussion, we record it before going on.

Corollary to the Proof of Theorem 6.1.8. A_6 is a simple group.

We return to the question of whether or not $n!/2$ is a square. As a matter of fact, it is *not* if $n > 2$. This can be shown as a consequence of the beautiful theorem in number theory (the so-called Bertrand Postulate), which asserts that for $m > 1$ there is always a prime between m and $2m$. Since we do not have this result at our disposal, we follow another road to show the simplicity of A_n for all $n \geq 5$.

We now prove this important theorem.

Theorem 6.1.9. For all $n \geq 5$ the group A_n is simple.

Proof. By Theorem 6.1.8 we may assume that $n \geq 6$. The center of A_n for $n > 3$ is merely (e). (Prove!) Since A_n is generated by the 3-cycles, if $\sigma \neq e$ is in A_n, then, for some 3-cycle τ, $\sigma\tau \neq \tau\sigma$.

Suppose that $N \neq (e)$ is a normal subgroup of A_n and that $\sigma \neq e$ is in N. Thus, for some 3-cycle τ, $\sigma\tau \neq \tau\sigma$, which is to say, $\sigma\tau\sigma^{-1}\tau^{-1} \neq e$. Because N is normal in A_n, the element $\tau\sigma^{-1}\tau^{-1}$ is in N, hence $\sigma\tau\sigma^{-1}\tau^{-1}$ is also in N. Since τ is a 3-cycle, so must $\sigma\tau\sigma^{-1}$ also be a 3-cycle. Thus N contains the product of two 3-cycles, and this product is not e. These two 3-cycles involve at most six letters, so can be considered as sitting in A_6 which, since $n \geq 6$, can be considered embedded isomorphically in A_n. (Prove!) But then $N \cap A_6 \neq (e)$ is a normal subgroup of A_6, so, by the Corollary above, $N \cap A_6 = A_6$. Therefore, N must contain a 3-cycle, and since all 3-cycles are conjugate in A_n (Lemma 6.1.6), N must contain all the 3-cycles in S_n. Since these 3-cycles generate A_n, we obtain that N is all of A_n, thereby proving the theorem. $\square$

There are many different proofs of Theorem 6.1.9—they usually involve showing that a normal subgroup of A_n must contain a 3-cycle—which are shorter and possibly easier than the one we gave. However, we like the bizarre twist in the proof given in that the whole affair boils down to the fact that 60 is not a square. We recommend to the reader to look at some other proofs of this very important theorem, especially in a book on group theory.

The A_n provide us with an infinite family of finite simple groups. There are several other infinite families of finite simple groups and 26 particular ones that do not belong to any infinite family. This determination of all finite simple groups, carried out in the 1960s and 1970s by a large number of group theorists, is one of the major achievements of twentieth-century mathematics.

Problems

*1. Prove that if $n > 2$, the center of S_n is (e).

*2. Prove that if $n > 3$, the center of A_n is (e).

3. What can you say about the cycle structures of the product of two 3-cycles?

4. If $m < n$, show that there is a subgroup of S_n isomorphic to S_m.

5. Show that an abelian group having no proper subgroups is cyclic of prime order.

6. How many conjugacy classes are there in S_6?

7. If the elements $a_1, a_2, \ldots, a_n$ generate the group G and b is a noncentral element of G, prove that $ba_i \neq a_i b$ for some i.

8. If $M \triangleleft N$ and $N \triangleleft G$, show that aMa^{-1} is normal in N for every $a \in G$.

9. If $M \triangleleft G$ and $N \triangleleft G$, show that MN is a normal subgroup of G.

10. If $n \geq 5$ is odd, show that the n-cycles generate A_n.

11. Show that the centralizer of $(12 \cdots k)$ in S_n has order $k(n-k)!$ and that $(12 \cdots k)$ has $n!/(k(n-k)!)$ conjugates in S_n.

12. In the proof of Theorem 6.1.8, show that $|MN| = |N|^2$.

2. FINITE FIELDS I

Our goal in this section and the next two is to get a complete description of all finite fields. What we shall show is that the multiplicative group of nonzero elements of a finite field is a cyclic group. This we do in this section. In the next two, the objectives will be to establish the existence and uniqueness of finite fields having p^n elements for any prime p and any positive integer n.

Some of the things that we are about to do already came up in the problem sets in group theory and field theory as hard problems. The techniques that we use come from group theory and field theory, with a little number theory thrown in.

We recall what the *Euler φ-function* is. We define the Euler φ-function by: $\varphi(1) = 1$ and, for $n > 1$, $\varphi(n)$ is the number of positive integers less than n and relatively prime to n.

We begin with a result in number theory whose proof, however, will exploit group theory. Before doing the general case, we do an example.

Let $n = 12$; then $\varphi(12) = 4$ for only 1, 5, 7, and 11 are less than 12 and relatively prime to 12. We compute $\varphi(d)$ for all the divisors of 12. We have: $\varphi(1) = 1$, $\varphi(2) = 1$, $\varphi(3) = 2$, $\varphi(4) = 2$, $\varphi(6) = 2$, and $\varphi(12) = 4$. Note that the sum of all $\varphi(d)$ over all the divisors of 12 is 12. This is no fluke but is a special case of

Theorem 6.2.1. If $n \geq 1$, then $\Sigma\varphi(d) = n$, where this sum runs over all divisors d of n.

Proof. Let G be a cyclic group of order n generated by the element a. If $d|n$, how many elements of G have order d? If $b = a^{n/d}$, then all the solutions in G of $x^d = e$ are the powers $e, b, b^2, \ldots, b^{d-1}$ of b. How many of these have order d? We claim, and leave to the reader, that b^r has order d if and only if r is relatively prime to d. So the number of elements of order d in G, for every divisor d of n, is $\varphi(d)$. Every element in G has order some divisor of n, so if we sum up the number of elements of order d—namely $\varphi(d)$—over all d dividing n, we account for each element of G once and only once. Hence $\Sigma\varphi(d) = n$ if we run over all the divisors d of n. The theorem is now proved. $\square$

In a finite cyclic group of order n the number of solutions of $x^d = e$, the unit element of G, is exactly d for every d that divides n. We used this fact in the proof of Theorem 6.2.1. We now prove a converse to this, getting thereby a criterion for cyclicity of a finite group.

> **Theorem 6.2.2.** Let G be a finite group of order n with the property that for every d that divides n there are at most d solutions of $x^d = e$ in G. Then G is a cyclic group.

Proof. Let $\psi(d)$ be the number of elements of G of order d. By hypothesis, if $a \in G$ is of order d, then all the solutions of $x^d = e$ are the distinct powers $e, a, a^2, \ldots, a^{d-1}$, of which number, $\varphi(d)$ are of order d. So if there is an element of order d in G, then $\psi(d) = \varphi(d)$. On the other hand, if there is no element in G of order d, then $\psi(d) = 0$. So for all $d|n$ we have that $\psi(d) \le \varphi(d)$. However, since every element of G has some order d that divides n we have that $\Sigma\psi(d) = n$, where this sum runs over all divisors d of n. But

$$n = \sum \psi(d) \le \sum \varphi(d) = n$$

since each $\psi(d) \le \varphi(d)$. This gives us that $\Sigma\psi(d) = \Sigma\varphi(d)$, which, together with $\psi(d) \le \varphi(d)$, forces $\psi(d) = \varphi(d)$ for every d that divides n. Thus, in particular, $\psi(n) = \varphi(n) \ge 1$. What does this tell us? *After all, $\psi(n)$ is the number of elements in G of order n, and since $\psi(n) \ge 1$ there must be an element a in G of order n.* Therefore, the elements $e, a, a^2, \ldots, a^{n-1}$ are all distinct and are n in number, so they must give all of G. Thus G is cyclic with a as generator, proving the theorem. $\square$

Is there any situation where we can be sure that the equation $x^d = e$ has at most d solutions in a given group? Certainly. If K^* is the group of nonzero elements of a field under multiplication, then the polynomial $x^n - 1$ has at most n roots in K^* by Theorem 5.6.2. So, if $G \subset K^*$ is a finite multiplicative subgroup of K^*, then the number of solutions of $x^d = 1$ in G is at most d for any positive integer d, so certainly for all d that divide the order of G. By Theorem 6.2.2 G must be a cyclic group.

We have proved

> **Theorem 6.2.3.** If K is a field and $K*$ is the group of nonzero elements of K under multiplication, then any finite subgroup of $K*$ is cyclic.

A very special case of Theorem 6.2.3, but at the moment the most important case for us, is

> **Theorem 6.2.4.** If K is a finite field, then $K*$ is a cyclic group.

Proof. $K*$ is a finite subgroup of itself, so, by Theorem 6.2.3, $K*$ is cyclic. □

A particular instance of Theorem 6.2.4 is of great importance in number theory, where it is known as the *existence of primitive roots mod p* for p a prime.

> **Theorem 6.2.5.** If p is a prime, then $\mathbb{Z}_p^*$ is a cyclic group.

Problems

1. If $a \in G$ has order d, prove that a^r also has order d if and only if r and d are relatively prime.

2. Find a cyclic generator (primitive root) for $\mathbb{Z}_{11}^*$.

3. Do Problem 2 for $\mathbb{Z}_{17}^*$.

4. Construct a field K having nine elements and find a cyclic generator for the group $K*$.

5. If p is a prime and $m = p^2$, then $\mathbb{Z}_m$ is not a field but the elements $\{[a] | (a, p) = 1\}$ form a group under the multiplication in $\mathbb{Z}_m$. Prove that this group is cyclic of order $p(p-1)$.

6. Determine all the finite subgroups of $\mathbb{C}*$, where $\mathbb{C}$ is the field of complex numbers.

In the rest of the problems here φ will be the Euler φ-function.

7. If p is a prime, show that $\varphi(p^n) = p^{n-1}(p-1)$.

8. If m and n are relatively prime positive integers, prove that

$$\varphi(mn) = \varphi(m)\varphi(n).$$

9. Using the result of Problems 7 and 8, find $\varphi(n)$ in terms of the factorization of n into prime power factors.

10. Prove that $\lim_{n \to \infty} \varphi(n) = \infty$.

3. FINITE FIELDS II: EXISTENCE

Let K be a finite field; then K must be of characteristic p, p a prime, and K contains $0, 1, 2, \ldots, p - 1$, the p multiples of the unit element 1 of K. So $K \supset \mathbb{Z}_p$, or, more precisely, K contains a field isomorphic to $\mathbb{Z}_p$. Since K is a vector space over $\mathbb{Z}_p$ and clearly is of finite dimension over $\mathbb{Z}_p$, if $[K : \mathbb{Z}_p] = n$, then K has p^n elements. This is true because, if $v_1, v_2, \ldots, v_n$ is a basis of K over $\mathbb{Z}_p$, then, for every distinct choice of $(\alpha_1, \alpha_2, \ldots, \alpha_n)$, where the α_i are in $\mathbb{Z}_p$, the elements

$$\alpha_1 v_1 + \alpha_2 v_2 + \cdots + \alpha_n v_n$$

are distinct. Thus, since we can pick $(\alpha_1, \alpha_2, \ldots, \alpha_n)$ in p^n ways, K has p^n elements.

Since K^*, the multiplicative group of nonzero elements of K, is a group of order $p^n - 1$ we have that $a^{m-1} = 1$, where $m = p^n$, for every a in K, hence $a^m = a$. Since this is also obviously true for $a = 0$, we have that $a^m = a$ for every a in K. Therefore, the polynomial $x^m - x$ in $\mathbb{Z}_p[x]$ has $m = p^n$ distinct roots in K, namely all the elements of K. Thus $x^m - x$ factors in $K[x]$ as

$$x^m - x = (x - a_1)(x - a_2) \cdots (x - a_m),$$

where $a_1, a_2, \ldots, a_m$ are the elements of K.

Everything we just said we already said, in more or less the same way, in Section 6 of Chapter 5. Since we wanted these results to be fresh in the reader's mind, we repeated this material here.

We summarize what we just did in

Theorem 6.3.1. Let K be a finite field of characteristic p, p a prime. Then K contains $m = p^n$ elements where $n = [K : \mathbb{Z}_p]$, and the polynomial $x^m - x$ in $\mathbb{Z}_p[x]$ splits into linear factors in $K[x]$ as

$$x^m - x = (x - a_1)(x - a_2) \cdots (x - a_m),$$

where $a_1, a_2, \ldots, a_m$ are the elements of K.

Two natural questions present themselves:

1. For what primes p and what integers n does there exist a field having p^n elements?
2. How many nonisomorphic fields are there having p^n elements?

We shall answer both questions in this section and the next. The answers will be

1. For any prime p and any positive integer n there exists a finite field having p^n elements.
2. Two finite fields having the same number of elements are isomorphic.

It is to these two results that we now address ourselves. First, we settle the question of the existence of finite fields. We begin with a general remark about irreducible polynomials.

Lemma 6.3.2. Let F be any field and suppose that $p(x)$ is an irreducible polynomial in $F[x]$. Suppose that $q(x)$ in $F[x]$ is such that in some extension field of F, $p(x)$ and $q(x)$ have a common root. Then $p(x)$ divides $q(x)$ in $F[x]$.

Proof. Suppose that $p(x)$ does not divide $q(x)$; since $p(x)$ is irreducible in $F[x]$, $p(x)$ and $q(x)$ must therefore be relatively prime in $F[x]$. Thus there are polynomials $u(x)$ and $v(x)$ in $F[x]$ such that

$$u(x)p(x) + v(x)q(x) = 1.$$

Suppose that the element a in some extension K of F is a root of both

$p(x)$ and $q(x)$; thus $p(a) = q(a) = 0$. But then $1 = u(a)p(a) + v(a)q(a) = 0$, a contradiction. So we get that $p(x)$ divides $q(x)$ in $F[x]$. $\square$

Note that we can actually prove a little more, namely

> **Corollary.** If $f(x)$ and $g(x)$ in $F[x]$ are not relatively prime in $K[x]$, where K is an extension of F, then they are not relatively prime in $F[x]$.

Let F be a field of characteristic $p \neq 0$. We claim that the polynomial $f(x) = x^m - x$, where $m = p^n$, *cannot* have a multiple root in any extension field K of F. Do you remember what a multiple root of a polynomial is? We refresh your memory. If $g(x)$ in $F[x]$ and if K is an extension field of F, then a in K is a *multiple* root of $g(x)$ if $g(x) = (x - a)^2 q(x)$ for some $q(x)$ in $K[x]$.

We return to the polynomial $f(x) = x^m - x$ above. Since $f(x) = x(x^{m-1} - 1)$ and 0 is not a root of $x^{m-1} - 1$, it is clearly true that 0 is a simple (i.e., not multiple) root of $f(x)$. Suppose that $\alpha \in K$, $K \supset F$, is a root of $f(x)$; thus $\alpha^m = \alpha$. If $y = x - \alpha$, then

$$f(y) = y^m - y = (x - \alpha)^m - (x - \alpha) = x^m - \alpha^m - (x - \alpha)$$

(since we are in characteristic $p \neq 0$ and $m = p^n$)

$$= x^m - x \text{ (because } \alpha^m = \alpha) = f(x).$$

So

$$f(x) = f(y) = y^m - y = (x - \alpha)^m - (x - \alpha)$$

$$= (x - \alpha)\big((x - \alpha)^{m-1} - 1\big),$$

and clearly this is divisible by $x - \alpha$ only to the first power, since $x - \alpha$ does not divide $(x - \alpha)^{m-1} - 1$. So α is not a multiple root of $f(x)$.

We have proved

> **Theorem 6.3.3.** If $n > 0$, then $f(x) = x^m - x$, where $m = p^n$, has no multiple roots in any field of characteristic p.

We should add a word to the proof above to nail down the statement of Theorem 6.3.3 as we gave it. Any field of characteristic $p \neq 0$ is an extension of $\mathbb{Z}_p$, and the polynomial $f(x)$ is in $\mathbb{Z}_p[x]$. So the argument above, with K any field of characteristic p and $F = \mathbb{Z}_p$, proves the theorem in its given form.

We have exactly what we need to prove the important

Theorem 6.3.4. For any prime p and any positive integer n there exists a finite field having p^n elements.

Proof. Consider the polynomial $x^m - x$ in $\mathbb{Z}_p[x]$, where $m = p^n$. By Theorem 5.6.6 there exists a finite extension K of $\mathbb{Z}_p$ such that in $K[x]$ the polynomial $x^m - x$ factors as

$$x^m - x = (x - a_1)(x - a_2) \cdots (x - a_m),$$

where $a_1, a_2, \ldots, a_m$ are in K. By Theorem 6.3.3, $x^m - x$ does not have any multiple roots in K, hence the elements $a_1, a_2, \ldots, a_m$ are $m = p^n$ distinct elements. We also know that $a_1, a_2, \ldots, a_m$ are *all* the roots of $x^m - x$ in K, since $x^m - x$ is of degree m.

Let $A = \{ a \in K | a^m = a \}$; as we just saw, A has m distinct elements. We claim that A is a field. If $a, b \in A$, then $a^m = a$ and $b^m = b$, hence $(ab)^m = a^m b^m = ab$, thus $ab \in A$. Because we are in characteristic $p \neq 0$ and $m = p^n$, $(a + b)^m = a^m + b^m = a + b$, hence $a + b$ is in A.

Since A is a finite subset of a field and is closed with respect to sum and product, A must be a subfield of K. Since A has $m = p^n$ elements, A is thus the field whose existence was asserted in the statement of the theorem. With this the theorem is proved. $\square$

Problems

*1. Give the details of the proof of the Corollary to Lemma 6.3.2.

The next two problems are a repeat of ones given earlier in the book.

2. If $f(x) = a_0 x^n + a_1 x^{n-1} + \cdots + a_n$ is in $F[x]$, let $f'(x)$ be the *formal derivative* of $f(x)$ defined by the following equation: $f'(x) = n a_0 x^{n-1} + (n-1) a_1 x^{n-2} + \cdots + (n-i) a_i x^{n-i-1} + \cdots + a_{n-1}$.

Prove that:

(a) $(f(x) + g(x))' = f'(x) + g'(x)$

(b) $(f(x)g(x))' = f'(x)g(x) + f(x)g'(x)$ for all $f(x)$ and $g(x)$ in $F[x]$.

*3. Prove that $f(x)$ in $F[x]$ has a multiple root in some extension of F if and only if $f(x)$ and $f'(x)$ are not relatively prime.

4. If $f(x) = x^n - x$ is in $F[x]$, prove that $f(x)$ does not have a multiple root in any extension of F if F is either of characteristic 0 or of characteristic $p \neq 0$, where p does *not* divide $n - 1$.

5. Use the result of Problem 4 to give another proof of Theorem 6.3.3.

6. If F is a field of characteristic $p \neq 0$, construct a polynomial with multiple roots of the form $x^n - x$, where $p|(n - 1)$.

7. If K is a field having p^n elements, show that for every m that divides n there is a subfield of K having p^m elements.

4.   FINITE FIELDS III: UNIQUENESS

Now that we know that finite fields exist having p^n elements, for any prime p and any positive integer n, we might ask: How many finite fields are there with p^n elements? For this to make any sense at all, what we are really asking is: How many distinct nonisomorphic fields are there with p^n elements? The answer to this is short and sweet: one. We shall show here that any two finite fields having the same number of elements are isomorphic.

Let K and L be two finite fields having p^n elements. Thus K and L are both vector spaces of dimension n over $\mathbb{Z}_p$. As such, K and L are isomorphic as *vector spaces*. On the other hand, K^* and L^* are both cyclic groups of order $p^n - 1$ by Theorem 6.2.4; hence K^* and L^* are isomorphic as *multiplicative groups*. One would imagine that one could put these two isomorphisms together to prove that K and L are isomorphic as *fields*. But it just isn't so. The proof does not take this direction at all. But the finiteness of K and L together with these two isomorphisms (of two structures carried by K and L) do *suggest* that, perhaps, K and L are isomorphic as fields. This is indeed the case, as we now proceed to show.

We begin with

Lemma 6.4.1. If $q(x)$ in $\mathbb{Z}_p[x]$ is irreducible of degree n, then $q(x)|(x^m - 1)$, where $m = p^n$.

Proof. By Theorem 4.5.11 the ideal $(q(x))$ of $\mathbb{Z}_p[x]$ generated by $q(x)$ is a maximal ideal of $\mathbb{Z}_p[x]$ since $q(x)$ is irreducible in $\mathbb{Z}_p[x]$. Let $A = \mathbb{Z}_p[x]/(q(x))$; by Theorem 4.4.3, A is a field of degree n over $\mathbb{Z}_p$, hence has p^n elements. Therefore, $u^m = u$ for every element u in A.

Let $a = x + (q(x))$ be the coset of x in $A = \mathbb{Z}_p[x]/(q(x))$; thus $q(a) = 0$ and $q(x)$ is the minimal polynomial for a over $\mathbb{Z}_p$. Since a is in A, $a^m = a$, so a is seen as a root of the polynomial $x^m - x$, where $m = p^n$. Thus $x^m - x$ and $q(x)$ have a common root in A. By Lemma 6.3.2 we have that $q(x)|(x^m - x)$. $\square$

We are now in a position to prove the main result of this section.

Theorem 6.4.2. If K and L are finite fields having the same number of elements, then K and L are isomorphic fields.

Proof. Suppose that K and L have p^n elements. By Theorem 6.2.4, L^* is a cyclic group generated, say, by the element b in L. Then, certainly, $\mathbb{Z}_p(b)$—the field obtained by adjoining b to $\mathbb{Z}_p$—is all of L. Since $[L : \mathbb{Z}_p] = n$, by Theorem 5.3.2 b is algebraic over $\mathbb{Z}_p$ of degree n, with $n = \deg(q(x))$, where $q(x)$ is the minimal polynomial in $\mathbb{Z}_p[x]$ for b, and is irreducible in $\mathbb{Z}_p[x]$.

The mapping $\psi : \mathbb{Z}_p(x) \to L = Z_p(b)$ defined by $\psi(f(x)) = f(b)$ is a homomorphism of $\mathbb{Z}_p[x]$ onto L with kernel $(q(x))$, the ideal of $\mathbb{Z}_p[x]$ generated by $q(x)$. So $L \simeq \mathbb{Z}_p[x]/(q(x))$.

Because $q(x)$ is irreducible in $Z_p[x]$ of degree n, by Lemma 6.4.1 $q(x)$ must divide $x^m - x$, where $m = p^n$. However, by Lemma 6.3.1, the polynomial $x^m - x$ factors in $K[x]$ as

$$x^m - x = (x - a_1)(x - a_2) \cdots (x - a_m),$$

where $a_1, a_2, \ldots, a_m$ are all the elements of K. Thus $q(x)$ divides $(x - a_1)(x - a_2) \cdots (x - a_m)$. By the Corollary to Theorem 4.5.10, $q(x)$ cannot be relatively prime to all the $x - a_i$ in $K[x]$, hence for some j, $q(x)$ and $x - a_j$ have a common factor of degree at least 1. In short, $x - a_j$ must divide $q(x)$ in $K[x]$, so $q(x) = (x - a_j)h(x)$ for some $h(x)$ in $K[x]$. Therefore, $q(a_j) = 0$.

Since $q(x)$ is irreducible in $\mathbb{Z}_p[x]$ and a_j is a root of $q(x)$, $q(x)$ must be the minimal polynomial for a_j in $\mathbb{Z}_p[x]$. Thus $\mathbb{Z}_p(a_j) \simeq \mathbb{Z}_p[x]/(q(x)) \simeq L$. This tells us, among other things, that we have $[\mathbb{Z}_p(a_j) : \mathbb{Z}_p] = n$, and since $\mathbb{Z}_p(a_j) \subset K$ and $[K : \mathbb{Z}_p] = n$ we conclude that $\mathbb{Z}_p(a_j) = K$. Therefore, $K = \mathbb{Z}_p(a_j) \simeq L$. Thus we get the result that we are after, namely, that K and L are isomorphic fields. This proves the theorem. $\square$

Combining Theorems 6.3.4 and 6.4.2, we have

> **Theorem 6.4.3.**   For any prime p and any positive integer n there exists, up to isomorphism, one and only one field having p^n elements.

5.   CYCLOTOMIC POLYNOMIALS

Let $\mathbb{C}$ be the field of complex numbers. As a consequence of De Moivre's Theorem the complex number $\theta_n = \cos 2\pi/n + i \sin 2\pi/n$ satisfies $\theta_n^n = 1$ and $\theta_n^m \neq 1$ if $0 < m < n$. We called θ_n a *primitive* nth *root of unity*. The other primitive nth roots of unity are

$$\theta_n^{\,k} = \cos\left(\frac{2\pi k}{n}\right) + i \sin\left(\frac{2\pi k}{n}\right),$$

where $(k, n) = 1$ and $1 \leq k < n$.

Clearly, θ_n satisfies the polynomial $x^n - 1$ in $\mathbb{Q}[x]$, where $\mathbb{Q}$ is the field of rational numbers. We want to find the minimal (monic) polynomial for θ_n over $\mathbb{Q}$.

We define a sequence of polynomials inductively. At first glance they might not seem relevant to the question of finding the minimal polynomial for θ_n over $\mathbb{Q}$. It will turn out that these polynomials are highly relevant to that question for, as we shall prove later, the polynomial $\phi_n(x)$ that we are about to introduce is a monic polynomial with integer

coefficients, is irreducible over $\mathbb{Q}$, and, moreover, $\phi_n(\theta_n) = 0$. This will tell us that $\phi_n(x)$ is the desired monic minimal polynomial for θ_n over $\mathbb{Q}$.
We now go about the business of defining these polynomials.

Definition. The polynomials $\phi_n(x)$ are defined inductively by:
(a) $\phi_1(x) = x - 1$.
(b) If $n > 1$, then $\phi_n(x) = (x^n - 1)/\prod\phi_d(x)$, where in the product in the denominator d runs over all the divisors of n except for n itself.
These polynomials are called the *cyclotomic polynomials* and $\phi_n(x)$ is called the nth *cyclotomic polynomial*.

At the moment it is not obvious that the $\phi_n(x)$ so defined are even polynomials, nor do we, as yet, have a clue as to the nature of the coefficients of these polynomials $\phi_n(x)$. All this will come in due time. But first we want to look at some early examples.

EXAMPLES

1. $\phi_2(x) = (x^2 - 1)/\phi_1(x) = (x^2 - 1)/(x - 1) = x + 1$.

2. $\phi_3(x) = (x^3 - 1)/\phi_1(x) = (x^3 - 1)/(x - 1) = x^2 + x + 1$.

3. $\phi_4(x) = (x^4 - 1)/(\phi_1(x)\phi_2(x)) = (x^4 - 1)/((x - 1)(x + 1)) = (x^4 - 1)/(x^2 - 1) = x^2 + 1$.

4. $\phi_5(x) = (x^5 - 1)/\phi_1(x) = (x^5 - 1)/(x - 1) = x^4 + x^3 + x^2 + x + 1$.

5. $\phi_6(x) = \dfrac{x^6 - 1}{\phi_1(x)\phi_2(x)\phi_3(x)}$

$$= \dfrac{x^6 - 1}{(x - 1)(x + 1)(x^2 + x + 1)}$$

$$= \dfrac{x^3 + 1}{x + 1} = x^2 - x + 1.$$

We notice a few things about the polynomials above:

1. They are all monic polynomials with integer coefficients.
2. The degree of $\phi_n(x)$ is $\varphi(n)$, where φ is the Euler φ-function, for $1 \le n \le 6$. (Check this out.)
3. Each $\phi_n(x)$, for $1 \le n \le 6$, is irreducible in $\mathbb{Q}(x)$. (Verify!)
4. For $1 \le n \le 6$, θ_n is a root of $\phi_n(x)$. (Verify!)

These few cases give us a hint as to what the general story might be for all $\phi_n(x)$. A hint, yes, but only a hint. To establish these desired properties for $\phi_n(x)$ will take some work.

To gain some further insight into these polynomials, we consider a particular case, one in which $n = p^m$, where p is a prime. To avoid cumbersome subscripts, we shall denote $\phi_n(x)$ by $\psi^{(m)}(x)$, where $n = p^m$. The prime p will be kept fixed in the discussion. We shall obtain explicit formulas for the $\psi^{(m)}(x)$'s and determine their basic properties. However, the method we use will not be applicable to the general case of $\phi_n(x)$. To study the general situation will require a wider and deeper set of techniques than those needed for $\psi^{(m)}(x)$.

We note one simple example. If p is a prime, the only divisor of p that is not p itself is 1. From the definition of $\phi_p(x) = \psi^{(1)}(x)$ we have that

$$\psi^{(1)}(x) = \phi_p(x) = \frac{x^p - 1}{x - 1} = x^{p-1} + \cdots + x + 1.$$

Note that in studying the Eisenstein Criterion we showed that this polynomial is irreducible in $\mathbb{Q}(x)$.

What can we say for the higher $\psi^{(m)}(x)$'s?

Lemma 6.5.1. For all $m \ge 1$,

$$\psi^{(m)}(x) = \frac{x^{p^m} - 1}{x^{p^{m-1}} - 1} = 1 + x^{p^{m-1}} + x^{2p^{m-1}} + \cdots + x^{(p-1)p^{m-1}}.$$

Proof. We go by induction on m.

If $m = 1$, we showed above that $\psi^{(1)}(x) = (x^p - 1)/(x - 1) = 1 + x + x^2 + \cdots + x^{p-1}$, so the lemma is true in this case.

Suppose that $\psi^{(r)} = (x^{p^r} - 1)/(x^{p^{r-1}} - 1)$ for all $r < m$. Consider $\psi^{(m)}(x)$; Since the only proper divisors of p^m are $1, p, p^2, \ldots, p^{m-1}$, from the definition of $\psi^{(m)}(x)$ we have that

$$\psi^{(m)}(x) = \frac{x^{p^m} - 1}{(x - 1)\psi^{(1)}(x) \cdots \psi^{(m-1)}(x)}.$$

By induction, $\psi^{(r)}(x) = (x^{p^r} - 1)/(x^{p^{r-1}} - 1)$ for $r < m$, hence

$$(x - 1)\psi^{(1)}(x) \cdots \psi^{(m-1)}(x)$$

$$= (x - 1)\frac{x^p - 1}{x - 1}\frac{x^{p^2} - 1}{x^p - 1} \cdots \frac{x^{p^{m-1}} - 1}{x^{p^{m-2}} - 1} = x^{p^{m-1}} - 1.$$

But then

$$\psi^{(m)}(x) = \frac{(x^{p^m} - 1)}{x^{p^{m-1}} - 1}$$

completing the induction and proving the lemma. $\square$

Note here that

$$\psi^{(m)}(x) = \frac{x^{p^m} - 1}{x^{p^{m-1}} - 1} = 1 + x^{p^{m-1}} + \cdots + x^{(p-1)p^{m-1}}$$

is a monic polynomial with integer coefficients. Its degree is clearly $p^{m-1}(p - 1)$, which is indeed $\varphi(p^m)$. Finally, if θ is a primitive p^mth root of unity, then $\theta^{p^m} = 1$, but $\theta^{p^{m-1}} \neq 1$, hence $\psi^{(m)}(\theta) = 0$; so θ is a root of $\psi^{(m)}(x)$. The final thing we want to know is: Is $\psi^{(m)}(x)$ irreducible over $\mathbb{Q}$?

Note that

$$\psi^{(m)}(x) = 1 + x^{p^{m-1}} + \cdots + x^{(p-1)p^{m-1}} = \psi^{(1)}(x^{p^{m-1}})$$

and we know that $\psi^{(1)}(x)$ is irreducible in $\mathbb{Q}[x]$. We shall use the Eisenstein Criterion to prove that $\psi^{(m)}(x)$ is irreducible in $\mathbb{Q}[x]$.

We digress for a moment. If $f(x)$ and $g(x)$ are two polynomials with integer coefficients, we define $f(x) \equiv g(x) \bmod p$ if $f(x) = g(x) + pr(x)$, where $r(x)$ is a polynomial with integer coefficients. This is equivalent to saying that the corresponding coefficients of $f(x)$ and $g(x)$ are congruent mod p. Expanding $(f(x) + g(x))^p$ by the binomial theorem, and using that all the binomial coefficients are divisible by p, since p is a prime, we arrive at $(f(x) + g(x))^p \equiv f(x)^p + g(x)^p \bmod p$.

Given $f(x) = a_0 x^n + a_1 x^{n-1} + \cdots + a_n$, where the a_i are integers, then, by the above,

$$f(x)^p = \left(a_0 x^n + a_1 x^{n-1} + \cdots + a_n\right)^p \equiv a_0^p x^{np} + a_1^p x^{np} + \cdots + a_n^p$$

$$\equiv a_0 x^{np} + a_1 x^{(n-1)p} + \cdots + a_n \quad \bmod p,$$

the latter congruence being a consequence of Fermat's Theorem (the Corollary to Theorem 2.4.8). Since $f(x^p) = a_0 x^{np} + a_1^{(n-1)p} + \cdots + a_n$, we obtain that

$$f(x^p) \equiv f(x)^p \bmod p.$$

Iterating what we just did, we arrive at

$$f(x^{p^k}) \equiv f(x)^{p^k} \bmod p$$

for all nonnegative k.

We return to our $\psi^{(m)}(x)$. Since $\psi^{(m)}(x) = \psi^{(1)}(x^{p^{m-1}})$ we have, from the discussion above, that $\psi^{(m)}(x) \equiv \psi^{(1)}(x^{p^{m-1}}) \bmod p$. Therefore,

$$\psi^{(1)}(x+1)^{p^{m-1}} = \left(\frac{(x+1)^p - 1}{(x+1) - 1}\right)^{p^{m-1}} = \left(\frac{(x+1)^p - 1}{x}\right)^{p^{m-1}}$$

$$= \left(x^{p-1} + px^{p-2} + \frac{p(p-1)}{2}x^{p-3} + \cdots + \frac{p(p-1)}{2}x + p\right)^{p^{m-1}}$$

$$\equiv \psi^{(1)}(x)^{p^{m-1}(p-1)} \bmod p \equiv \psi^{(m)}(x+1) \bmod p.$$

This tells us that

$$\psi^{(m)}(x+1) = x^{p^{m-1}(p-1)} + pr(x),$$

where $r(x)$ is a polynomial with integer coefficients. So all the coefficients of $\psi^{(m)}(x + 1)$, with the exception of its leading coefficient 1, are divisible by p. If we knew for some reason that the constant term of $h(x) = \psi^{(m)}(x + 1)$ was not divisible by p^2, we could apply the Eisenstein Criterion to show that $h(x)$ is irreducible. But what is the constant term of $h(x) = \psi^{(m)}(x + 1)$? It is merely $h(0) = \psi^{(m)}(1)$, which, from the explicit form of $\psi^{(m)}(x + 1)$ that we found four paragraphs earlier, is exactly p. Thus $h(x)$ is irreducible in $Q[x]$, that is, $\psi^{(m)}(x + 1)$ is irreducible in $\mathbb{Q}[x]$. But this immediately implies that $\psi^{(m)}(x)$ is irreducible in $\mathbb{Q}[x]$.

Summarizing, we have proved

> **Theorem 6.5.2.** For $n = p^m$, where p is any prime and m any nonnegative integer, the polynomial $\phi_n(x)$ is irreducible in $\mathbb{Q}[x]$.

As we pointed out earlier, this is a very special case of the theorem we shall soon prove; namely, that $\phi_n(x)$ is irreducible for all positive integers n. Moreover, the result and proof of Theorem 6.5.2 play no role in the proof of the general proposition that $\phi_n(x)$ is irreducible in $\mathbb{Q}[x]$. But because of the result in Theorem 6.5.2 and the explicit form of $\phi_n(x)$ when $n = p^m$, we do get a pretty good idea of what should be true in general. We now proceed to the discussion of the irreducibility of $\phi_n(x)$ for general n.

> **Theorem 6.5.3.** For every integer $n \geq 1$,
>
> $$\phi_n(x) = (x - \theta^{(1)}) \cdots (x - \theta^{(\varphi(n))}),$$
>
> where $\theta^{(1)}, \theta^{(2)}, \ldots, \theta^{(\varphi(n))}$ are the $\varphi(n)$ distinct primitive nth roots of unity.

Proof. We proceed by induction on n.

If $n = 1$, then $\phi_1(x) = x - 1$, and since 1 is the only first root of unity, the result is certainly correct in this case.

Suppose that result is true for all $m < n$. Thus, if $d \mid n$ and $d \neq n$, then, by the induction, $\phi_d(x) = (x - \theta_d^{(1)}) \cdots (x - \theta_d^{(\varphi(d))})$, where the $\theta_d^{(i)}$ are the primitive d th roots of unity. Now

$$x^n - 1 = (x - \zeta_1)(x - \zeta_2) \cdots (x - \zeta_n),$$

where the ζ_i run over *all* n th roots of unity. Separating out the primitive n th roots of unity in this product, we obtain

$$x^n - 1 = (x - \theta^{(1)}) \cdots (x - \theta^{(\varphi(n))}) v(x),$$

where $v(x)$ is the product of all the other $x - \zeta_i$; thus by our induction hypothesis $v(x)$ is the product of the $\phi_d(x)$ over all divisors d of n with the exception of $d = n$. Thus, since

$$\phi_n(x) = \frac{(x^n - 1)}{v(x)} = \frac{(x - \theta^{(1)}) \cdots (x - \theta^{(\varphi(n))}) v(x)}{v(x)}$$

$$= (x - \theta^{(1)})(x - \theta^{(2)}) \cdots (x - \theta^{(\varphi(n))}),$$

we have proved the result claimed in the theorem. $\square$

From the form of $\phi_n(x)$ in Theorem 6.5.3 we immediately see that $\phi_n(x)$ is a monic polynomial in $\mathbb{C}[x]$ of degree $\varphi(n)$. Knowing this, we prove that, in fact, the coefficients of $\phi_n(x)$ are integers. Why is this true? Proceeding by induction on n, we may assume this to be the case if $d|n$ and $d \neq n$. Therefore, if $v(x)$ denotes the polynomial used in the proof of Theorem 6.5.3, then $(x^n - 1)/v(x) = \phi_n(x) \in \mathbb{C}[x]$, hence $v(x)|x^n - 1$ in $\mathbb{C}[x]$. But, by the long-division process, dividing the monic polynomial $v(x)$ with integer coefficients into $x^n - 1$ leads us to a monic polynomial with integer coefficients (and no remainder, since $v(x)|(x^n - 1)$ in $\mathbb{C}[x]$). Thus $(x^n - 1)/v(x) = \phi_n(x)$ is a monic polynomial with integer coefficients. As we saw, its degree is $\varphi(n)$. Thus

Theorem 6.5.4. For every positive integer n the polynomial $\phi_n(x)$ is a monic polynomial with integer coefficients of degree $\varphi(n)$, where φ is the Euler φ-function.

Knowing that $\phi_n(x)$ is a polynomial, we can see that its degree is $\varphi(n)$ in yet another way. From $\phi_n(x) = (x^n - 1)/v(x)$, using induction on n, $\deg(\phi_n(x)) = n - \deg(v(x)) = n - \Sigma\varphi(d)$, the sum over all divisors d of n other than $d = n$, from the form of $v(x)$. Invoking the result of Theorem 6.2.1, $n - \Sigma\varphi(d) = \varphi(n)$, where again this sum is over all $d|n$, $d \neq n$. We thus obtain that $\deg(\phi_n(x)) = \varphi(n)$.

The result we are about to prove is without question one of the most basic ones about cyclotomic polynomials.

Theorem 6.5.5.   For every positive integer n the polynomial $\phi_n(x)$ is irreducible in $Q[x]$

Proof.   Let $f(x)$ in $Q[x]$ be an irreducible polynomial such that $f(x)|\phi_n(x)$. Thus $\phi_n(x) = f(x)g(x)$ for some $g(x)$ in $Q[x]$. By Gauss' Lemma we may assume that both $f(x)$ and $g(x)$ are monic polynomials with integer coefficients, thus are in $Z[x]$. Our objective is to show that $\phi_n(x) = f(x)$; if this were the case, then, since $f(x)$ is irreducible in $Q[x]$, we would have that $\phi_n(x)$ is irreducible in $Q[x]$.

Since $\phi_n(x)$ has no multiple roots, $f(x)$ and $g(x)$ must be relatively prime. Let p be a prime number such that p does not divide n. If θ is a root of $f(x)$, it is then a root of $\phi_n(x)$, hence by Theorem 6.5.3 θ is a primitive nth root of unity. Because p is relatively prime to n, θ^p is also a primitive nth root of unity, thus, by Theorem 6.5.3, θ^p is a root of $\phi_n(x)$. We therefore have that $0 = \phi_n(\theta^p) = f(\theta^p)g(\theta^p)$, from which we deduce that either $f(\theta^p) = 0$ or $g(\theta^p) = 0$.

Our aim is to show that $f(\theta^p) = 0$. Suppose not; then $g(\theta^p) = 0$, hence θ is a root of $g(x^p)$. Because θ is also a root of the irreducible polynomial $f(x)$, by Lemma 6.3.2 we obtain that $f(x)|g(x^p)$. As we saw in the course of the proof of Theorem 6.5.2, $g(x^p) \equiv g(x)^p \bmod p$.

Let J be the ideal in Z generated by p; by the Corollary to Theorem 4.6.2, $Z[x]/J[x] \simeq Z_p[x]$, which means that reducing the coefficients of any polynomial mod p is a homomorphism of $Z[x]$ onto $Z_p[x]$.

Since all the polynomials $\phi_n(x)$, $v(x)$, $f(x)$, and $g(x)$ are in $Z[x]$, if $\bar\phi_n(x)$, $\bar v(x)$, $\bar f(x)$, and $\bar g(x)$ are their images in $Z_p[x]$, all the relations among them are preserved going mod p. Thus we have the relations $x^n - 1 = \bar\phi_n(x)\bar v(x)$, $\bar\phi_n(x) = \bar f(x)\bar g(x)$ and $\bar f(x)|\bar g(x^p) = \bar g(x)^p$.

Therefore, $\bar f(x)$ and $\bar g(x)$ have a common root, a, in some extension K of Z_p. Now $x^n - 1 = \bar\phi_n(x)\bar v(x) = \bar f(x)\bar g(x)$, hence a, as a root of both $\bar f(x)$ and $\bar g(x)$ is a multiple root of $x^n - 1$. But the formal derivative $(x^n - 1)'$ of $x^n - 1$ is $nx^{n-1} \neq 0$, since p does not divide n; therefore, $(x^n - 1)'$ is relatively prime to $x^n - 1$. By the result of Problem 3 of Section 3 the polynomial $x^n - 1$ *cannot* have a multiple root. With this contradiction arrived at from the assumption that θ^p was not a root of $f(x)$, we conclude that whenever θ is a root of $f(x)$, then so must θ^p be one, for any prime p that does not divide n.

Repeating this argument, we arrive at θ^r is a root of $f(x)$ for every integer r that is relatively prime to n. But θ, as a root of $f(x)$, is a root of $\phi_n(x)$, so is a primitive nth root of unity. Thus θ^r is also a primitive nth root of unity for every r relatively prime to n. By running over all r that are relatively prime to n, we pick up every primitive nth root of unity as some such θ^r. *Thus all the primitive nth roots of unity are roots of* $f(x)$. By Theorem 6.5.3 we see that $\phi_n(x) = f(x)$, hence $\phi_n(x)$ is irreducible in $\mathbb{Q}[x]$. $\square$

It may strike the reader as artificial and unnatural to have resorted to the passage mod p to carry out the proof of the irreducibility of a polynomial with rational coefficients. In fact, it may very well be artificial and unnatural. As far as we know, no proof of the irreducibility of $\phi_n(x)$ has ever been given staying completely in $\mathbb{Q}[x]$ and not going mod p. It would be esthetically satisfying to have such a proof. On the other hand, this is not the only instance where a result is proved by passing to a related subsidiary system. Many theorems in number theory —about the ordinary integers—have proofs that exploit the integers mod p.

Because $\phi_n(x)$ is a monic polynomial with integer coefficients which is irreducible in $\mathbb{Q}[x]$, and since θ_n, the primitive nth root of unity, is a root of $\phi_n(x)$, we have

> **Theorem 6.5.6.** $\phi_n(x)$ is the minimal polynomial in $\mathbb{Q}[x]$ for the primitive nth roots of unity.

Problems

1. Verify that the first six cyclotomic polynomials are irreducible in $\mathbb{Q}[x]$ by a direct frontal attack.

2. Write down the explicit forms of:
   (a) $\phi_{10}(x)$.
   (b) $\phi_{15}(x)$.
   (c) $\phi_{20}(x)$.

3. If $(x^m - 1)|(x^n - 1)$, prove that $m|n$.

4. If $a > 1$ is an integer and $(a^m - 1)|(a^n - 1)$, prove that $m|n$.

5. If K is a finite extension of $\mathbb{Q}$, the field of rational numbers, prove
   that there is only a finite number of roots of unity in K (**Hint:** Use
   the result of Problem 10 of Section 2, together with Theorem 6.5.6.)

6. LIOUVILLE'S CRITERION

Recall that a complex number is said to be *algebraic of degree n* if it is
the root of a polynomial of degree n over $\mathbb{Q}$, the field of rational
numbers, and is not the root of any such polynomial of degree lower
than n. In the terms used in Chapter 5, an algebraic number is a
complex number algebraic over $\mathbb{Q}$.

A complex number that is not algebraic is called *transcendental*. Some
familiar numbers, such as e, π, e^{π}, and many others, are known to be
transcendental. Others, equally familiar, such as $e + \pi$, $e\pi$, and π^e, are
suspected of being transcendental but, to date, this aspect of their nature
is still open.

The French mathematician Joseph Liouville (1809–1882) gave a
criterion that any algebraic number of degree n must satisfy, This
criterion gives us a condition that limits the extent to which a real
algebraic number of degree n can be approximated by rational numbers.
This criterion is of such a nature that we can easily construct real
numbers that violate it for every $n > 1$. Any such number will then have
to be transcendental. In this way we shall be able to produce transcendental numbers at will. However, none of the familiar numbers is
such that its transcendence can be proved using Liouville's Criterion.

In this section of the book we present this result of Liouville. It is a
surprisingly simple and elementary result to prove. This takes nothing
away from the result; in our opinion it greatly enhances it.

> **Theorem 6.6.1 (Liouville).** Let a be an algebraic number of
> degree $n \geq 2$ (i.e., is, a is algebraic but not rational). Then
> there exists a positive constant c (which depends only on a)
> such that for all integers u, v with $v > 0$, $|a - u/v| > c/v^n$.

Proof. Let a be a root of the polynomial $f(x)$ of degree n in $\mathbb{Q}[x]$,
where $\mathbb{Q}$ is the field of rational numbers. By clearing of denominators in

the coefficients of $f(x)$, we may assume that $f(x) = r_0 x^n + r_1 x^{n-1} + \cdots + r_n$, where all the r_i are integers and where $r_0 > 0$.

Since the polynomial $f(x)$ is irreducible of degree n it has n distinct roots $a = a_1, a_2, \ldots, a_n$ in $\mathbf{C}$, the field of complex numbers. Therefore, $f(x)$ factors over $\mathbf{C}$ as $f(x) = r_0(x - a)(x - a_2) \cdots (x - a_n)$. Let u, v be integers with $v > 0$; then

$$f\left(\frac{u}{v}\right) = \frac{r_0 u^n}{v^n} + \frac{r_1 u^{n-1}}{v^{n-1}} + \cdots + \frac{r_{n-1} u}{v} + r_n,$$

hence

$$v^n f\left(\frac{u}{v}\right) = r_0 u^n + r_1 u^{n-1} v + \cdots + r_{n-1} u v^{n-1} + r_n v^n$$

is an integer. Moreover, since $f(x)$ is irreducible in $Q[x]$ of degree $n \geq 2$, $f(x)$ has no rational roots, so $v^n f(u/v)$ is a nonzero integer, whence $|v^n f(u/v)| \geq 1$. Using the factored form of $f(x)$, we have that

$$f\left(\frac{u}{v}\right) = r_0\left(\left(\frac{u}{v}\right) - a\right)\left(\left(\frac{u}{v}\right) - a_2\right) \cdots \left(\left(\frac{u}{v}\right) - a_n\right),$$

hence

$$\left|\left(\frac{u}{v}\right) - a\right| = \frac{|f(u/v)|}{r_0|(u/v) - a_2| \cdots |(u/v) - a_n|}$$

$$= \frac{v^n |f(u/v)|}{r_0 v^n |(u/v) - a_2| \cdots |(u/v) - a_n|}$$

$$\geq \frac{1}{r_0 v^n |(u/v) - a_2| \cdots |(u/v) - a_n|}.$$

Let s be the largest of $|a|, |a_2|, \ldots, |a_n|$. We divide the argument according as $|u/v| > 2s$ or $|u/v| \leq 2s$. If $|u/v| > 2s$, then, by the triangle inequality, $|a - (u/v)| \geq |u/v| - |a| > 2s - s = s$, and, since $v \geq 1$, $|a - (u/v)| > s/v^n$.

On the other hand, if $|u/v| \leq 2s$, then, again by the triangle inequality, $|a_i - (u/v)| \leq |a_i| + |u/v| \leq s + 2s = 3s$. Therefore,

$$t = \left| a_2 - \left(\frac{u}{v} \right) \right| \cdots \left| a_n - \left(\frac{u}{v} \right) \right| \leq (3s)^{n-1},$$

so that $1/t \geq 1/(3s)^{n-1} = 1/(3^{n-1}s^{n-1})$. Going back to the inequality above that $|a - (u/v)| \geq 1/[r_0 v^n |a_2 - (u/v)| \cdots |a_n - (u/v)|]$, we have that $|a - (u/v)| \geq 1/(r_0 3^{n-1} s^{n-1} v^n)$. These numbers $r_0, 3^{n-1}, s^{n-1}$ are determined once and for all by a and its minimal polynomial $f(x)$ and *do not depend on u or v*. If we let $b = 1/(r_0 3^{n-1} s^{n-1})$, then $b > 0$ and $|a - (u/v)| > b/v^n$. This covers the second case, where $|u/v| \leq 2s$.

If c is a positive number smaller than both b and s, we have from the discussion that $|a - u/v| > c/v^n$ for all integers u, v, where $v > 0$, thereby proving the theorem. $\square$

Let's see the particulars of the proof for the particular case $a = \sqrt{2}$. The minimal polynomial for a in $\mathbb{Q}[x]$ is $f(x) = (x - a)(x + a)$, so $a = a_1$ and $-a = a_2$. So if u and v are integers, and $v > 0$, then

$$v^2 f\left(\frac{u}{v} \right) = v^2 \left(\left(\frac{u}{v} \right)^2 - a^2 \right) = v^2 \left(\left(\frac{u}{v} \right)^2 - 2 \right) = u^2 - 2v^2 \neq 0,$$

an integer. So $|v^2 f(u/v)| \geq 1 \geq 1/v^2$. The s above is the larger of $|\sqrt{2}|$ and $|-\sqrt{2}|$; that is, $s = \sqrt{2}$. Also, the b above is $1/(3^{2-1}(\sqrt{2})^{2-1}) = 1/(3\sqrt{2})$, so if c is any positive number less than $1/(3\sqrt{2})$, then $|\sqrt{2} - u/v| > c/v^2$.

What the theorem says is the following: Any algebraic real number has rational numbers as close as we like to it (this is true for all real numbers), but if this algebraic real number a is of degree $n \geq 2$, there are restrictions on how finely we can approximate a by rational numbers. These restrictions are the ones imposed by Liouville's Theorem.

How do we use this result to produce transcendental numbers? All we need do is to produce a real number τ, say, such that whatever positive integer n may be, and whatever positive c we choose, we can find a pair of integers u, v, with $v > 0$ such that $|\tau - u/v| < c/v^n$. We can find such a τ easily by writing down an infinite decimal involving 0's and 1's, where we make the 0's spread out between the 1's very rapidly. For instance, $\tau = 0.10100100000010\ldots010\ldots$, where the 0's between successive 1's go like $m!$ is a number that violates Liouville's Criterion for every $n > 0$. (See Problem 3, Section 3). Thus this number τ is transcendental.

We could, of course, use other wide spreads of 0's between the 1's—m^m, $(m!)^2$, and so on—to produce hordes of transcendental numbers. Also, instead of using just 1's, we could use any of the nine nonzero digits to obtain more transcendental numbers. We leave the verification that the numbers of the sort we described do not satisfy Liouville's Criterion for any positive integer n and any positive c.

We can use the transcendental number τ and the variants of it we described to prove a famous result due to Cantor. This result says that there is a 1-1 correspondence between all the real numbers and its subset of real transcendental numbers. In other words, in some sense, there are as many transcendental reals as there are reals. We give a brief sketch of how we carry it out, leaving the details to the reader.

First, it is easy to construct a 1-1 mapping of the reals onto those reals strictly between 0 and 1 (try to find such a mapping). This is also true for the real transcendental numbers and those of them strictly between 0 and 1. Let the first set be denoted by A and the second one by B. We shall construct a 1-1 mapping of A into B. This will be enough to finish our task.

Given any number in A, we can represent it as an infinite decimal $a_1a_2 \cdots a_n \ldots$, where the a_i fall between 0 and 9. (We now wave our hands a little, being a little bit inaccurate. The reader should try to tighten up the argument.) Define the mapping f from A to B by $f(a_1a_2 \cdots a_n \ldots) = 0.a_10a_200a_3000000a_4 \ldots$; by the Liouville Criterion, except for a small set of $a_1, a_2, \ldots, a_n \ldots$, the numbers $0.1a_10a_200a_3000000a_4 \ldots$ are transcendental. The f we wrote down then provides us with the required mapping.

One final word about the kind of approximation of algebraic numbers by rationals expressed in Theorem 6.6.1. There we have that if a is real algebraic of degree $n \geq 2$, then $|a - u/v| > c/v^n$ for some appropriate positive c. If we could decrease the n to $|a - u/v| > c/v^m$ for $m < n$ and some suitable c (depending on a and m), we would get an even sharper result. In 1955 the (then) young English mathematician K. F. Roth proved the powerful result that effectively we could cut the n down to 2. His exact result is: If a is algebraic of degree $n \geq 2$, then for every real number $r > 2$ there exists a positive constant c, depending on a and r, such that $|a - u/v| > c/v^r$ for all but a finite number of fractions u/v.

7.　THE IRRATIONALITY OF π

As we indicated earlier, Lindemann in 1882 proved that π is a transcendental number. In particular, from this result of Lindemann it follows that π is irrational. We shall not prove the transcendence of π here—it would require a rather long detour—but we will, at least, prove that π is irrational. The very nice proof that we give of this fact is due to I. Niven; it appeared in his paper "A Simple Proof That π Is Irrational," which was published in the *Bulletin of the American Mathematical Society*, vol. 53 (1947), p. 509. To follow Niven's proof only requires some material from a standard first-year calculus course.

We begin with

Lemma 6.7.1. If u is a real number, then $\lim_{n \to \infty} u^n/n! = 0$.

Proof. If u is any real number, then e^u is a well-defined real number and $e^u = 1 + u + u^2/2! + u^3/3! + \cdots + u^n/n! + \cdots$. The series $1 + u + u^2/2! + \cdots + u^n/n! + \cdots$ converges to e^u; since this series converges, its nth term must go to 0. Thus $\lim_{n \to \infty} u^n/n! = 0$. $\square$

We now present Niven's proof of the irrationality of π.

Theorem 6.7.2. π is an irrational number.

Proof. Suppose that π is rational; then $\pi = a/b$, where a and b are positive integers.

We introduce a polynomial, based on the assumption that $\pi = a/b$, for every integer $n > 0$, whose properties will lead us to the desired conclusion. The basic properties of this polynomial will hold for all positive n. The strategy of the proof is to make a judicious choice of n at the appropriate stage of the proof.

Let $f(x) = x^n(a - bx)^n/n!$, where $\pi = a/b$. This is a polynomial of degree $2n$ with rational coefficients. Expanding it out, we obtain that

$$f(x) = \frac{a_0 x^n + a_1 x^{n+1} + \cdots + a_n x^{2n}}{n!},$$

where

$$a_0 = a^n, a_1 = -na^{n-1}b, \ldots, a_i = \frac{(-1)^i n!}{i!(n-i)!} a^{n-i} b^i, \ldots, a_n = (-1)^n b^n$$

are integers.

We denote the ith derivative of $f(x)$ with respect to x by the usual notation $f^{(i)}(x)$, understanding $f^{(0)}(x)$ to mean $f(x)$ itself.

We first note a symmetry property of $f(x)$, namely, that $f(x) = f(\pi - x)$. To see this, note that $f(x) = (b^n/n!)x^n(\pi - x)^n$, from whose form it is clear that $f(x) = f(\pi - x)$. Since this holds for $f(x)$, it is easy to see, from the chain rule for differentiation, that $f^{(i)}(x) = (-1)^i f^{(i)}(\pi - x)$.

This statement about $f(x)$ and all its derivatives allows us to conclude that for the statements that we make about the nature of all the $f^{(i)}(0)$, there are appropriate statements about all the $f^{(i)}(\pi)$.

We shall be interested in the value of $f^{(i)}(0)$, and $f^{(i)}(\pi)$, for all nonnegative i. Note that from the expanded form of $f(x)$ given above we easily obtain that $f^{(i)}(0)$ is merely $i!$ times the coefficient of x^i of the polynomial $f(x)$. This immediately implies, since the lowest power of x appearing in $f(x)$ is the nth, that $f^{(i)}(0) = 0$ if $i < n$. For $i \geq n$ we obtain that $f^{(i)}(0) = i!a_{i-n}/n!$; since $i \geq n$, $i!/n!$ is an integer, and as we pointed out above, a_{i-n} is also an integer; therefore $f^{(i)}(0)$ *is an integer for all nonnegative integers i.* Since $f^{(i)}(\pi) = (-1)^i f(0)$, we have that $f^{(i)}(\pi)$ is an integer for all nonnegative integers i.

We introduce an auxiliary function

$$F(x) = f(x) - f^{(2)}(x) + \cdots + (-1)^n f^{(2n)}(x).$$

Since $f^{(m)}(x) = 0$ if $m > 2n$, we see that

$$\frac{d^2 F}{dx^2} = F''(x) = f^{(2)}(x) - f^{(4)}(x) + \cdots + (-1)^n f^{(2n)}(x)$$

$$= -F(x) + f(x).$$

Therefore,

$$\frac{d}{dx}(F'(x)\sin x - F(x)\cos x) = F''(x)\sin x + F'(x)\cos x$$

$$- F'(x)\cos x + F(x)\sin x$$

$$= (F''(x) + F(x))\sin x = f(x)\sin x.$$

From this we conclude that

$$\int_0^\pi f(x)\sin x\,dx = [F'(x)\sin x - F(x)\cos x]_0^\pi$$

$$= (F'(\pi)\sin\pi - F(\pi)\cos\pi) - (F'(0)\sin 0 - F(0)\cos 0)$$

$$= F(\pi) + F(0).$$

But from the form of $F(x)$ above and the fact that all $f^{(i)}(0)$ and $f^{(i)}(\pi)$ are integers, we conclude that $F(\pi) + F(0)$ is an integer. *Thus $\int_0^\pi f(x)\sin x\,dx$ is an integer. This statement about $\int_0^\pi f(x)\sin x\,dx$ is true for any integer $n > 0$ whatsoever.* We now want to choose n cleverly enough to make sure that the statement "$\int_0^\pi f(x)\sin x\,dx$ is an integer" cannot possibly be true.

We carry out an estimate on $\int_0^\pi f(x)\sin x\,dx$. For $0 < x < \pi$ the function $f(x) = x^n(a - bx)^n/n! \le \pi^n a^n/n!$ (since $a > 0$), and also $0 < \sin x \le 1$. Thus $0 < \int_0^\pi f(x)\sin x\,dx < \int_0^\pi \pi^n a^n/n!\,dx = \pi^{n+1}a^n/n!$.

Let $u = \pi a$; then, by Lemma 6.7.1, $\lim_{n \to \infty} u^n/n! = 0$, so *if we pick n large enough, we can make sure that $u^n/n! < 1/\pi$*, hence $\pi^{n+1}a^n/n! = \pi u^n/n! < 1$. But then $\int_0^\pi f(x)\sin x\,dx$ is trapped strictly between 0 and 1. But, by what we have shown, $\int_0^\pi f(x)\sin x\,dx$ is an integer. Since there is no integer strictly between 0 and 1, we have reached a contradiction. Thus the premise that π is rational was false. Therefore, π is irrational. This completes the proof of the theorem. $\square$

INDEX

289